A MULTIDISCIPLINARY
INTRODUCTION TO
INFORMATION
SECURITY

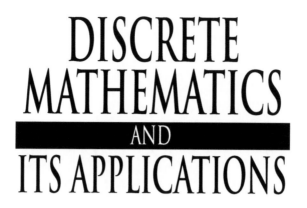

DISCRETE MATHEMATICS AND ITS APPLICATIONS

Series Editor

Kenneth H. Rosen, Ph.D.

DISCRETE MATHEMATICS AND ITS APPLICATIONS

Series Editor KENNETH H. ROSEN

A MULTIDISCIPLINARY INTRODUCTION TO INFORMATION SECURITY

Stig F. Mjølsnes

Norwegian University of Science & Technology
Trondheim

CRC Press
Taylor & Francis Group
Boca Raton   London   New York

CRC Press is an imprint of the
Taylor & Francis Group, an **informa** business

A CHAPMAN & HALL BOOK

CRC Press
Taylor & Francis Group
6000 Broken Sound Parkway NW, Suite 300
Boca Raton, FL 33487-2742

First issued in paperback 2017

ISBN 13: 978-1-138-11213-1 (pbk)
ISBN 13: 978-1-4200-8590-7 (hbk)

Preface

Information security is a truly multidisciplinary field of study, ranging from the methods of pure mathematics through computer and telecommunication sciences to social sciences. The intention of this multi-authored book is to offer an introduction to a wide set of topics in ICT information security, privacy, and safety. Certainly, the aim has not been to present a complete treatment of this vast and expanding area of practical and theoretical knowledge. Rather, my hope is that the selected range of topics presented here may attract a wider audience of students and professionals than would each specialized topic by itself.

Some of the information security topics contained in this book may be familiar turf for the reader already. However, the reader will likely find some new relevant topics presented here that can enhance his or her professional knowledge and competence, or serve as an attractive starting point for further reading and in-depth studies. For instance, the book may provide an entrance and a guide to seek out more specialized courses available at universities or inspire further work in projects and assignments.

The start of this collection of information security topics goes back to a master-level continuing education course that I organized in 2005, where more than 10 professors and researchers contributed from six different departments at the Norwegian University of Science and Technology. The topics included cryptography, hardware security, software security, communication and network security, intrusion detection systems, access policy and control, risk and vulnerability analysis, and security technology management. The compendium of the lecturers' presentations then grew into a book initiative taken on by the Norwegian University of Science and Technology's Strategic Research Programme Committee for Information Security, which I was heading. And more authors were asked to contribute with hot topics as this project grew.

The topics and chapters in this book could have been ordered by many reasonable and acceptable principles. I chose to start with the basic components of hardware and algorithms, move toward integration and systems, and end with a chapter on human factors in these systems.

Many interdependencies and some overlap exist between the chapters, of course, for instance, the electronic hardware realizations in Chapter 1 and the public-key algorithms in Chapter 2, so a total linear sequence of the chapters in this respect has not been possible to set. The index at the back of the book is meant to be a helpful guide to find all chapters and locations that deal with a specific keyword or problem issue.

The book's cover drawing and all chapter front drawings are made especially for this book by Hannah Mjølsnes. This process went something like this. First, I tried to explain in simple words what the chapter was about, and then she made some pencil sketches of illustration ideas that we discussed. At a later stage, she worked out the complete illustrations on drawing paper, digitized these by scanning, and finally did the necessary postprocessing of the digital images for use in this book.

Acknowledgments

I wish to thank all the contributing authors for their effort and positive attitude toward this book project. Some of this sure took a while! Thank you to all the technical reviewers for your time and valuable recommendations to improve the text. None mentioned none forgotten. Thanks to PhD-students Anton Stolbunov and Mauritz Panggebean who assisted me in typesetting the manuscripts and bibliographies from authors not versed in LaTeX. A big hug to fine art student Hannah Mjølsnes for all the amusing and diverting artwork you made for this book.

I am most grateful to the CRC representative Robert B. Stern who accepted this book project back then, for his patient and considerate guidance and excellent recommendations throughout the years. I would also like to thank the rest of the people I communicated with in the publication process at Taylor and Francis Group; Amber Donley, Scott Hayes, Jim McGovern, Katy Smith, all your requests and advice were clear, professional and understandable.

Stig Frode Mjølsnes

Contributors

Einar Johan Aas
Department of Electronics and Telecommunications
Norwegian University of Science and Technology, Trondheim
einar.j.aas@ntnu.no

Eirik Albrechtsen
Department of Industrial Economy and Technology Management
Norwegian University of Science and Technology, Trondheim
eirik.albrechtsen@iot.ntnu.no

Jan Arild Audestad
Department of Telematics
Norwegian University of Science and Technology, Trondheim
Gjøvik University College, Gjøvik
audestad@item.ntnu.no

Martin Eian
Department of Telematics
Norwegian University of Science and Technology, Trondheim
martin.eian@item.ntnu.no

Danilo Gligoroski
Department of Telematics
Norwegian University of Science and Technology, Trondheim
danilog@item.ntnu.no

Stein Haugen
Department of Production and Quality Engineering
Norwegian University of Science and Technology, Trondheim
stein.haugen@ntnu.no

Dag Roar Hjelme
Department of Electronics and Telecommunications
Norwegian University of Science and Technology, Trondheim
dag.hjelme@iet.ntnu.no

Jan Hovden
Department of Industrial Economy and Technology Management
Norwegian University of Science and Technology, Trondheim
jan.hovden@iot.ntnu.no

Martin Gilje Jaatun
Department of Software Engineering, Safety and Security
SINTEF ICT, Trondheim
martin.g.jaatun@sintef.no

Jostein Jensen
Department of Software Engineering, Safety and Security
SINTEF ICT, Trondheim
jostein.jensen@sintef.no

Per Gunnar Kjeldsberg
Department of Electronics and Telecommunications
Norwegian University of Science and Technology, Trondheim
per.gunnar.kjeldsberg@iet.ntnu.no

Svein Johan Knapskog
Department of Telematics
Norwegian University of Science and Technology, Trondheim
svein.knapskog@item.ntnu.no

Lars Lydersen
Department of Electronics and Telecommunications
Norwegian University of Science and Technology, Trondheim
lars.lydersen@gmail.com

Vadim Makarov
University Graduate Center, Kjeller
makarov@vad1.com

Per Håkon Meland
Department of Software Engineering, Safety and Security
SINTEF ICT, Trondheim
per.h.meland@sintef.no

Stig Frode Mjølsnes
Department of Telematics
Norwegian University of Science and Technology, Trondheim
stig.mjolsnes@item.ntnu.no

Sverre Olaf Smalø
Department of Mathematical Sciences
Norwegian University of Science and Technology, Trondheim
sverresm@math.ntnu.no

Inger Anne Tøndel
Department of Software Engineering, Safety and Security
SINTEF ICT, Trondheim
inger.a.tondel@sintef.no

Svein Yngvar Willassen
Department of Telematics
Norwegian University of Science and Technology, Trondheim
svein@willassen.no

List of Figures

List of Tables

Contents

10 A Lightweight Approach to Secure Software Engineering · 183

M. G. Jaatun, J. Jensen, P. H. Meland and I. A. Tøndel

1

Introduction

S. F. Mjølsnes

Department of Telematics, NTNU

CONTENTS

1.1 Motivation

The recent two decades has seen a rapid shift in business and governance of telecommunications from mail and plain old telephone services to computer communications services, foremost e-mail, web transactions, and mobile devices. New information and communication technologies certainly create new conditions for the best structures and optimal organizations of our work and of our leisure time too. The challenge of finding the best practice and use of a multitude of emerging information and communications technologies will be with us for many decades to come. This fundamental transition to new ways of offering services by digital information exchange cuts through all sectors in our society and creates a demand for a variety of new specialists and expertise. ICT information security, privacy, and resilience are in the front row of the challenges we must find solutions for. As a result, industry and governments express growing needs for knowledge and experts that can propose viable solutions and make working implementations.

1.2   What Is Information Security?

We achieve security by being able to defend against attacks that might happen to us. Here "us" is the asset to be protected in the system of concern; "attacks that might happen" are the threats to these assets. An attack is an *intentional* act originated by somebody in the environment. In many situations, it is natural to make a distinction between outside and inside threats of attack. Faults, mistakes, accidents, on the other hand, are unintentional events that can lead to bad consequences, but there is no cunning behind such incidents. Figure 1.1 makes a distinction between the unwanted events of a system.

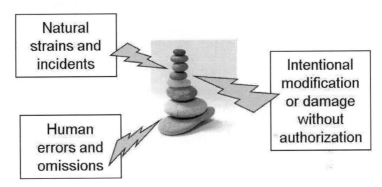

FIGURE 1.1
Categories of unwanted events that can happen to a system.

The adverbs *secure* and *sure* originate from the same root in English. In Norwegian and German language and possibly others, the very same word (Norwegian: *sikker*; German: *sicher*) is used in both contexts. Are you secure? and Are you sure? both translate to "Er du sikker?" in Norwegian. So what are we sure about? The state of being secure against a threat or danger means that we have taken precautions somehow, and that we are sure that we can manage if the threat materializes.

We can follow several strategies to secure a system against attacks :

- install measures that *prevent* the attack to be possible to carry out,

- install measures that will *detect and respond* when the attack is carried out,

- install measures that will *correct and recover* from damaged to normal operation.

Data have hardly any value in themself, when separated from their use or potential use. Data become valuable when someone depends on or wants the information represented by the data. Typically, data produced, processed, and stored in a laptop computer used for business purposes are worth much more than the actual laptop hardware for the owner. This is why information security focuses on how to protect the most expensive asset, namely, the information processed in computers and communicated on the networks.

Donn Parker [1] relates an entertaining story that makes this point crystal clear to us, even though the events must have taken place more than 30 years ago.

A former military officer was hired as data security officer in a company running a large central computing facility. The first thing he took on was to install an automatic carbon dioxide fire suppression system. So far so good, but what about the people that often worked in the computer facility? His solution was a big brass lever mounted above the main computer console that should be used by the chief operator to open and close the fire extinguisher gas manually. Unfortunately, it turned out that it was not possible to see the total area from the main console position. Then our security officer hung some gas masks on the wall, to be available for people that happened to be in that zone. Someone then pointed out that a normal gas mask filters out toxic materials but cannot aid when the air lacks oxygen! Anyway, the security officer continued his methods of physical security by ordering 20 mm caliber cannons to be placed at the entrance of the computing facility, securing against possible raids from the university area across the street. Then the man was fired.[1]

1.3   Some Basic Concepts

1.3.1   The Communication Perspective

A formal definition of information is closely related to the theory of communication, so we will derive some of the basic concepts of information security by studying the simple model of communication: two independent entities, a sender, Alice, and a recipient, Bob, connected by means of a communication channel of some sort. The standard model in information security assumes that the communication channel is outside the control of Alice and Bob, therefore available to attack.

The basic concepts of confidentiality, authenticity, and availability in information security will be introduced by listing a number of possible attack threats to the channel and the messages sent. However, do not think that this will remain the exhaustive list of possible threats in a more refined scenario. Your list of possible attacks will grow as you put on more detail to your scenario.

---

[1] Pun intended.

Confidentiality

The scenario here is simply that Alice will confide a private message to Bob using the communication channel connecting them.

The first threat that we identify is that a recipient other than Bob is able to receive the message intended for Bob only. This is often referred to as *wiretapping, eavesdropping, or interception* of the message. Another threat is that a nonintended recipient is able to determine that Alice is sending a message to Bob, at which time the communication is sent, the length of the message, and other details about the transmission process. This is referred to as *traffic analysis*. A third threat is that a nonintended recipient is able to read the content of the message, even though it may be encrypted. This is called *cryptanalysis*. A fourth threat is that Bob will break Alice's confidence assumption and reveal (parts of) the message. Bob is now an *adversary* of Alice. This latter threat might seem quite far fetched and dramatic, but if you think of it, this threat corresponds to many practical existing systems. Consider, for instance, the hot issue of the digital content and immaterial rights distribution problem in the music industry.

These are all security concerns of the sender, that is, the entity that releases information on the communication channel. If both Alice and Bob take on the *role* as senders in a two-way communication setting, then both will have to be concerned with the confidentiality of the information that is communicated. We can distinguish between *unidirectional* and *bidirectional* communication between the two *parties*. And more general, there may be many Bobs; that is, Alice will confide a private message to a set of intended recipients, by *multicast* or *broadcast* communication.

Authenticity

The starting scenario for the concept of authenticity is that Bob receives a message on the communication channel that claims to be from Alice, and where the content of the message is a public statement from Alice, all correct and complete as she sent it.

Authenticity is a quite rich and wide but also a slippery concept that is hard to pin down in a simple manner or capture in a single definition that will fit all. Our scenario here explicitly states that the message is meant for the public, thereby creating independence from the foregoing scenario of confidentiality. The worrying is primarily on Bob, the recipient, in this authenticity scenario.

One threat is that an attacker is able to generate a message and send to Bob on behalf of Alice. These kinds of threats are often referred to as *masquerading, spoofing,* or *impostor* attacks. Another threat is that the attacker resends a copy of an earlier recorded genuine message sent by Alice. This is referred to as a *replay* attack. A third threat is that Bob really receives a message from Alice, but parts of the message may be deleted, permuted, substituted, or inserted before it reaches Bob. This can be referred to as a *modification, forgery*, or *integrity* attack.

We can attempt a more precise definition, for instance, of message integrity, to show the more slippery side of capturing the notion of authenticity:

> The property of *message integrity* relative to Alice is maintained at the receiving side if the message that Bob receives is the same as the message Alice sent.

Ponder this definition while considering the processing of the source messsage down the communication protocol stack of the transmitter, possibly some transcoding performed within the network, and finally the unpacking and local presentation at receiver. Now what exactly do we mean by "the same" here? It does not really help much to precision if we add a qualification like "if the essential content of the message is the same" and throw in some hefty hand-waving. Remember that Alice and Bob are in all likelihood not human beings, but some computing devices that need unequivocal instructions what to do.

Availability

An illustrative scenario for availability is where Alice will send an alarm message to Bob if and whenever an emergency situation for her occurs, so that Bob will respond by initiating a rescue operation.

This alarm scenario brings forth the property of timeliness of the communication. It is a real-time issue. As the scenario above goes, Alice wants the alarm message to be received with negligible delay after she sent it. A quite parallel scenario that shifts the interest of timeliness to Bob is the following: Alice is the guard of Bob's vault, and if and whenever a burglar enters then she will set off an alarm message to Bob, so that Bob will come to her assistance.

Whereas the primary object of the properties of confidentiality and authenticity is the information itself, the property of availability brings the focus onto the actual service provided and the threats of *service disruption*.

The first threat is that Bob never receives Alice's message. We can refer to this as a *deletion, blocking*, or *jamming* attack. More generally, the threat is that the message can be *delayed* in transmission, a *degradation of service*. Certainly, if the attacker is in full control of the communication channel, then there is not anything that can be done about this threat. We may be able to reassess the power of the attacker and weaken the assumption of the attacker. Or we have to rethink outside the box, for instance, by introducing a *broadcast channel* or several independent communication channels, which are harder for the attacker to fully control. Second, the attacker can generate *fake alarm* messages or even replay messages. The false alarm effect will likely be similar to the story of the shepherd boy that cried 'wolf' too many times, nobody will care to respond after experiencing of multiple false warnings. A third threat is that Bob becomes too busy responding to other alarms and will not have sufficient resources left for responding to Alice. This is often referred to as a *denial-of-service* attack if the triggered alarms are generated by the attacker.

Alice and Bob may agree on an *active off* scheme, where Bob will be alarmed if he fails to receive regular messages from Alice. A threat to this scheme will be that the intruder is able to continue to generate the regular messages or a denial-of-service.

1.3.2 The Shared Computer Perspective

Now let us look at information security concepts as seen from the perspective of sharing computer resources.

Access Control

The purpose of the access control of computer systems is twofold. First there is normally a need to distinguish between the authorized users and the non-users. Second, once the authorized users are recognized and accepted ("logged in") there are many good reasons for distinguishing between the authorized users. Sharing can be on many levels, such as hardware, input and output lines, storage, data and program files, and running processes. Access control mechanisms allow several users to share computer resources in an orderly fashion.

Here are some general concepts that are used in most computer access control systems. Each user is assigned a unique *user identity*. The user identity is associated with a set of *rights* to objects of the resource. Such objects may be files, folders, programs, input/output resources, or it may be more refined structure, as for instance in data base systems. *Ownership* is a user's special right to establish, dispense and revoke access rights for a given object. Typical access rights in operating systems are **create, read, write, execute, delete**. An access control list or *authorization list* for an object is a list of user identities and associated access rights for this object. The authorization list is normally itself an object, and must be carefully protected as a system object.

The first premise for correct operation of the access control system is that all access requests to objects must be mediated by the access control mechanism. Obviously, the access control system will not be effective if there exist "side doors and loop holes." The access control decides whether to grant an access request based on the authorization of the user and the *access policy* of the computer system. The second premise for correct operation is that the logged in user identity corresponds to the correct user. For this, we need a user authentication protocol based upon something the user remembers (e.g., a secret password, character string or sentence), and/or something the user holds (e.g., a token or smart card), and/or a biometric characteristic of the user (e.g., voice, fingerprint, face, DNA).

A distinction in design between the actual access enforcement mechanism and the description of the access policy will allow for greater flexibility in the application of the specific access control design. This is similar to the distinction made between a program specification and its implementation mechanism. The specification tells *what* to do, and the mechanism tells *how* it will be done. The advantages of this abstraction is

1. The security requirements and rules can be managed outside the particular enforcement mechanisms.

2. Policies can be compared without comparing the execution mechanisms.

3. The same access control mechanisms can be used with different security policy designs.

Information Classification

Information can be attributed with some cost or value in some sense or another. Confidential data can be classified in levels of sensitivity. Data in a company can be classified according to department or the purpose. Health care data can be classified according to privacy regulations and laws.

A *discretionary* access policy is characterized by letting the owner, at his or her discretion, set the access rights. This is how we introduced the access control system above. A *mandatory* access policy goes further by letting the access system itself impose the policy rules that apply without considering each individual user. This is where standard classification of objects and of users come in handy because this will make it easier to state the policy that the system will follow without considering the particular object instances and their content.

A typical fully ordered classification hierarchy is Unclassified, Confidential, Secret, and Top Secret. A typical partly ordered classification is the possible subsets of the set of company departments, e.g., the set {Financial, Personel, Marketing}.

Information Flow

Access control enforces that all references to objects are authorized. This can control creating, reading, writing, and deleting information objects by rejecting nonauthorized access. The implicit assumption in access control is that authorized users will behave according to the rules. *Flow control* mechanisms target how information "flow" from one object to others. Information flow control mechanism try to set and enforce rules for how the users can disseminate and merge information. For instance, a strong concept of *noninterference* has been proposed:

> One group of users, using a certain set of commands, is non-interfering with another group of users if what the first group does with those commands has no effect on what the second group of users can see [2].

We note that computing and storage hardware have become so inexpensive and easily available today that the impetus of the past of reducing cost by time-sharing of computer hardware is not there anymore. Nevertheless, we are increasingly sharing information over the internet, across organizational and national boundaries, for instance, using large, globalized web services within the all-encompassing notion of centralized cloud computing. Hence, we will continue to share computers because we will distribute and share information.

Observe that the notion of information flow control easily brings us back to where we started at in this section, namely, the communication perspective.

1.4   A Synopsis of the Topics

1.4.1   The Book Structure

The book contains a selected subset of topics from the wide field of information security problems and solutions related to information and communication technology. Each chapter presents a particular topic or problem area and ends with recommendations for further reading and interesting web sites that contain more in-depth information. Citations and references are listed at the end of each chapter. This should make it easier to read a chapter independently of other chapters.

The basic idea behind the book's ordering of chapters is to start with the basic building blocks and move toward the systems. This establishes a first approximation for the dependencies of the chapters. Nevertheless, there are interdependencies between chapters too, such as the public key algorithms presented in Chapter 3 and the description about how these algorithms can be implemented in electronic hardware in Chapter 2. Each of the Chapters 10 to 13 can certainly be read without a deep understanding of the earlier chapters. Chapter 14 on security management assumes familiarity with the content of Chapter 13 on risk assessment.

You will find a synopsis of the problems dealt with in each chapter below. The index in the back of the book can be used to find all chapters that refer to a key word or topic.

1.4.2   Security Electronics

We often make a distinction between the hardware and the software of a computing device. We can easily touch and inspect the hardware, such as the keyboard, the circuitry boards with electronic components inside the cabinet, the silicon patterns inside the integrated chips, and so on. Software is a more abstract concept. The software has to do with the flexibility of the operational behavior of the device, to which extent we are able to program the behaviour of the device. A general-purpose microprocessor is, in a sense, a realization of a universal machine in that it can execute any computation algorithm, bounded only by the memory resources it has available.

This means that a general-purpose microprocessor can be programmed to execute any security mechanism that we want, bounded only by execution time and memory space. However, this flexibility comes with a cost compared to special-purpose processors and circuitry. Hardware with reduced programming flexibility can result in less hardware cost, higher execution speed, and less energy consumption.

Chapter 2 explains how this special hardware realization can be done for the cryptographic algorithms of AES and RSA. AES is the current international standard for symmetric key crypto-algorithm. It is very likely that both your mobile SIM card and your credit card chip contain the AES, as well as your WiFi-enabled laptop computer. RSA is a public key crypto-algorithm that require arithmetic computations on very large integers. The CPU registries of general-purpose processors are much too short for these long integers, so special hardware can reduce the computation time to a great extent.

1.4.3 Public Key Cryptography

Chapter 3 explains the mathematical reasoning behind the concept of public key cryptography and shows the number-theoretic design of the RSA algorithm. First, the public and the private keys must be generated, starting out with two large random prime numbers. Then the public key is computed and used for the encryption. The amazing property is that this encryption key can be made publicly readable without harming the confidentiality of the encrypted message. Efficient decryption can only be made with input of a corresponding decryption key. This decryption key must be kept secret by the intended recipients of the message, hence the term private key. Naturally, the private key must be related to the public key somehow, but the security claim is that in practice it is not possible to compute the private key from the public key. The encryption key can be made public so that everybody can encrypt, but only the holder of the private key can decrypt. Chapter 3 also introduces some more concepts of public key cryptography, such as digital signatures and hash functions.

1.4.4 Hash Functions

Chapter 4 describes the concept of a cryptographic hash function and its required properties in depth. A hash function computes a one-way short digest of its input. It takes input values of any length and outputs a function value of short length, say, between 128 and 512 bits. Furthermore, the hash function must be oneway so that it is practically impossible to compute an unknown input value that corresponds to a known function value.

Hash functions are very important in a plethora of applications. They are standard tools in password protection, digital signature generation, message authentication coding, commitment protocols, file and string pattern recognition, pseudorandom number generation, and much more.

1.4.5 Quantum Cryptography

While the security of cryptographic algorithms are based on problems that are assumed to be computationally hard and infeasible to solve by computers, the security of quantum cryptography is based on the laws of quantum physics. Chapter 5 presents the working principles of quantum cryptography, and gives an example of a quantum cryptographic protocol and its implementation using technology available today. This is an exciting alternative approach to communication secrecy based on the laws of physics rather than the hard problems of algorithmics. The principles were proposed in the 1980s as mere theoretical possibilities working with single polarized photons, but quite soon people managed to realize the ideas. First over short air channels (only 30–50 cm) but soon with optical fibers over tens of kilometers distance. Today, it is possible to purchase industrial grade crypto-equipment based on quantum cryptography principles.

1.4.6 Cryptographic Protocols

A common language is essential to communication. A language is built on the alphabet of symbols, the syntax of acceptable words, and the grammar of sentences. The notion of a communication protocol can be considered analogous to a language. The communicating parties need to establish the set of possible messages (words) in the exchange, and the behavior of message exchange must be carried out according to *the protocol*, that is, the prior agreement of the communicating parties. Chapter 6 gives an introduction to the special problems that come with the use of *cryptographic primitives* in the communication process. The first three chapters of this book all presented special cryptographic primitives, such as AES, RSA, and SHA. Chapter 6 presents how these and other cryptographic transformations can be used in communication. Some very surprising communication problems can be solved with cryptography. For instance, two communicating parties can simulate the process of "flipping a fair coin" by a cryptographic protocol, something that is quite impossible without cryptographic means.

1.4.7 Public Key Infrastructure

Chapter 7 starts out with the problem of crypto-key distribution and explains how the use of public key certification can mitigate this. The concept of public key infrastructure involves a networked infrastructure of servers, Certification Authorities, that distribute *certified public keys*. This certification of key authenticity deals with the problem of which public key belongs to which network user. A Certification Authority will bind a public key with a network name and address, and certify this relation by issuing a digital signature. The resulting data object is called a public key certificate. The general vision of Public Key Infrastructure emerged from telecom industries, international standardization, and government administration needs. Over the years, many realization attempts of such an infrastructure have been fraught with both technological and practical difficulties, such as the naming problem, privacy concerns, incompatibility with existing business models and organization, and the impact of client/user assistance and support.

1.4.8 Wireless Network Access

The technology of telecommunications by electromagnetic radio waves has been developing for more than one hundred years now. When applied as a network access link solution it allows for a wireless and mobile communication terminal. Whereas wires are easily confined within a physical perimeter, such as inside a building, radio signals are not. Chapter 8 discusses the security problems of wireless network access. Passive listening to a radio-based communication cannot be detected, and active impostors can act from the network addresses of regular users without breaking physical barriers. In particular, the chapter presents the solutions developed in the IEEE 802.11 standard to protect the radio link between the client station and the network access point.

1.4.9 Mobile Security

Chapter 9 starts with the standardization of the GSM system which took place early in the 1980s, where the objective was to specify a common land mobile system for Europe. Later the ambition of GSM changed from Groupe Spécial Mobile to *Global* System for Mobile Communications, thereafter upgraded to a sweeping *Universal* Mobile Telephone System. Chapter 9 describes the start and the evolution of these systems, and the existing protection mechanisms of the radio access link.

Currently, we are "all" subscribers to the global mobile telecommunication system, probably the largest machine in the universe. The number of subscribers in the GSM/UMTS worldwide in 2010 is estimated to about 4,450,000,000, in other words more than half the world's total population are now online by wireless access. China is the largest with 700 million subscribers, and India is second with 500 million subscribers. And we are in the middle of the roll out of the 4G mobile systems, which promise wireless access rates in the range of 100 Mbps to 1 Gbps, thereby enabling, for instance, high-quality bidirectional video communication applications in a mobile setting. The now widely popular short message service (SMS), "texting," was only hesitantly put in the original standard while questioning the real need for this functionality. Texting is now a large industry of its own, with security applications such as providing one-time access codes to internet banking users.

The security mechanisms of GSM are focused on two goals: access control and the security of the radio channels between the mobile station and the access network. For access control, the subscriber must be identified and authenticated in order for the network operator to make correct accounting and billing. Moreover, the identity of the subscriber is concealed such that several calls cannot be traced to the same subscriber by radio channel eavesdropping. For communication security, the the radio link transmissions are kept confidential by encryption. The UMTS adapted the successful parts of the GSM security architecture and added a few more. The UMTS mobile station is also able to authenticate the network it is connecting to, and the radio link communication can be explicitly integrity protected. However, the problem of end-to-end security was an issue both during the development stages and in the standardization of UMTS, but this functionality was eventually dropped because of national security requirements and the problem of export controls for strong cryptography.

1.4.10 Software Security

The Lennon–McCartney tune *Fixing a Hole* comes to my mind when the problem of software security is raised:

```
I'm fixing a hole where the rain gets in
And stops my mind from wandering where it will go
I'm filling the cracks that ran through the door
And kept my mind from wandering where it will go
```

We download and install software security patches to our networked computers on a regular basis these days. By this practice, we mend "the holes" detected in the operating system, web browsers, and other programs that constitute the operational computer. Taken as a sewing metaphor immediately guide us to the question: why did the tailor leave those holes in the garment in the first place? This is an obvious challenge to the tailor, that is, the programmer. Well, one common response is that he overlooked the error in the cut somehow, an *implementation error*. Another response might be that he actually thought the design was supposed to be like that, a *design error*. In both cases, the tailor's proficiency must be questioned. Or even his guildhall!

Let us give an example of security software, that is, a security mechanism implemented and run in software. The example is an access control mechanism that requests a username and password, checks the input with an access control list, and allows entry to the application if the input data match the recorded reference in the access control list. The access control software module will accept all listed usernames input together with a matching password. Furthermore, it will reject all listed usernames input together with a non-matching password. So this will be working reliably and consistently for all listed usernames. Nevertheless, the access control module might still be totally insecure against active attacks.

Chapter 10 proposes a software engineering practice that considers the vulnerability to potential attacks and misuse of the software in the early stages, as well as throughout the development process.

1.4.11   ICT Security Evaluation

The engineering of a new idea often starts out with a rough sketch on paper or a whiteboard depicting boxes, lines, arrows, and the like, accompanied with some oral explanation and hand-waving. This might be a convincing first step, but as the saying goes, the proof of the pudding is in the eating, so the pudding has to made before the final word can be said. The abstract ideas and concepts for a novel security device or system have to be interpreted and realized into an actual working construction. The standard process of engineering goes through several stages. It starts out with a description of the structure and the functionality, often with increasing level of details in several rounds of refined specifications. These design specifications can be checked and tested by engineers with relevant experience using methods of science and mathematical rigor. Then, at some point, we must determine which of the available electronic and physical components that shall make up the parts of the physical tangible device. And we must compile and link the software and embed the executable machine code into the device. The overall process of design and implementation adds a lot of structure and functionality to the original starting point. How can we convince ourselves and others that the device will actually work as required and that the security properties are effective and appropriate?

Chapter 11, "ICT Security Evaluation," describes and explains an evaluation method that aims to assure that the security requirements are as claimed for a product or a system. The method is an international standard called *Common Criteria*. It recommends how to specify security requirements, how to structure the description of the *target of evaluation*, and the requirements of the evaluation process itself.

1.4.12   ICT and Forensic Science

The pervasiveness of computers and networks both in private and public use guarantees an abundance of digital evidence to the inquisitive questions: *Where were you? What happened there? Who did it?* Although acts of crime might primarily be directed at physical targets and people, they will leave digital tracks in one or more electronic devices "witnessing" the course of the incident.

Sherlock Holmes[2] closed his eyes and placed his elbows upon the arms

---

[2]Arthur Conan Doyle's fictional character that features in four novels and 56 short stories, covering a time period from about 1880 to 1914.

of his chair, with his finger-tips together. "The ideal reasoner," he re-marked, "would, when he had once been shown a single fact in all its bearings, deduce from it not only all the chain of events which led up to it but also all the results which would follow from it. As Cuvier could correctly describe a whole animal by the contemplation of a single bone, so the observer who has thoroughly understood one link in a series of incidents should be able to accurately state all the other ones, both before and after. [...] To carry the art, however, to its highest pitch, it is necessary that the reasoner should be able to utilise all the facts which have come to his knowledge; and this in itself implies, as you will readily see, a possession of all knowledge, which, even in these days of free education and encyclopaedias, is a somewhat rare accomplishment.[3]

The forensic principles used in solving the mysteries of Holmes remain in-spirational to us living in the information age. Here is another basic rule of Holmes.

I have no data yet. It is a capital mistake to theorise before one has data. Insensibly one begins to twist facts to suit theories, instead of theories to suit facts.

Chapter 12 examines technical approaches to investigation of crimes that involve electronic computing devices and digital communication in some way. The methods and knowledge of this investigative process and the tools needed for this particular type of evidence is often denoted as *digital* forensic science, or digital forensics for short. Digital forensics is the investigation process concerned with evidence gathered from the digital tracks. *Computational forensics* involves supporting all technical investigations and evidence analysis with computational instrumentation, thereby bringing in objective measurements rather than human assessment.

1.4.13 Risk Assessment

Murphy's law states that "If anything can go wrong, it will." This pessimistic prediction of the future is rather rudimental, and quite self-evident and brings little information.

A more detailed analysis wants to know what exactly can go wrong, when and how often will it fail, what may the causes be, and what will be the consequences and cost of the failure. One commonly used probabilistic definition states that the measure of risk is equal to the product of the probability of an incident happening and the expected loss caused by the incident. If loss can be measured as cost in units of money, then the comprehension of risk will be easy by the business community, where it will be directly compatible

---

[3]From *The Adventures of Sherlock Holmes.*

with the management processes of budgets, accounting, and investment analyses. Chapter 13 describes risk assessment as part of the larger activity of risk management.

Information security is mostly concerned with the risk of deliberate incidents that are caused by adversaries using computers and communications. These adversaries are not random processes caused by indifferent natural phenomena, but intentional acts with intelligent adaptive gaming behavior. The element of surprise makes it hard to base the risk assessment on experience only. Even the knowledge of precisely which attacks are feasible is hard to come by in complex ICT systems. And statistical data needed for good probability estimators of the potential bad events cannot be collected because fixes and precautions will be taken once it happens, and then the preconditions change. Nevertheless, the chapter proposes a systematic process to risk assessment, which is a much better alternative than calling defeat in the business and organizational management process.

1.4.14   The Human Factor

Chapter 14, "Information Security Management" discusses information security from the organizational and people point of view. What should we be concerned about when we are organizing people around the changing ICT technology in business companies, public organizations, and institutions. Computers that store, process, and communicate data are the core part of many businesses today, though it is the information that the equipment holds that is valuable. The span is vast, from a military organization commanding "need-to-know" principles to public institutions that should abide with "right-to-know" principles. The information security policies are obviously very different in those two scenarios; nevertheless, we will find that desktops and laptop computers, printers, and local area networks with switches and routers are parts of both of these organizations. This is so because the security functionality of the devices is built to accommodate a range of security policies. The theory says that the security policy description should be separated from the security functionality itself, to be added by configuration and programming, and by how the devices are put together and structured into a system. Obviously, there will be technical limits to realizing a specific security policy. However, the greater challenge to the organization is actually to describe and specify the security policy in sufficient detail to be useful.

The names of Alice and Bob are often used in the research community and papers to label the communicating parties in cryptographic protocols. Although the researchers are actually talking about communicating computers, not human beings, the anthropomorphic narrative makes it easier to convey intuitively the security goals of the information security problem dealt with. The attacker is often named Malice and is depicted as an evil being, and Alice has a private key that only she "knows." We can conclude that the security requirements and policy must originate in the human factor, not in the indifferent machines.

1.5 Further Reading and Web Sites

First of all, you can browse the chapters of this book and search their bibliographies for further reading on your topic of interest.

There are many textbooks directed to the special fields of *information security*. You can find books specialized on computer security, communication security, internet security, database security, operating systems security, cryptography, security management, software security, browser security, physical security, security models, and the list could go on filling a whole page or more. One comprehensive masters level book that I have experience with is Ref. [3], although the students find it perhaps a bit tedious on technical details at times. *Security Engineering* [4] is a comprehensive book with lots of examples from technology practice. The first edition is available online at `http://www.cl.cam.ac.uk/~rja14/book.html`.

The Code Book [5] is an entertaining popular science book about the evolution of secret writing and coding from early history up to modern cryptography. *Handbook of Applied Cryptography* [6] is a comprehensive coverage of the field that is fully available online at `http://www.cacr.math.uwaterloo.ca/hac/`. The International Association for Cryptologic Research provides a web site [7] for ongoing cryptology research activities.

Check out the openBSD site at `http://www.openbsd.org/` if you want to get serious about operating system security. Cipher newsletter [8] of the IEEE Computer Society's TC on Security and Privacy reports on current events. The NIST Computer Security Resource Center web site at `http://csrc.nist.gov/` carries a wealth of publications on security standards and best practices.

Bibliography

[1] D. B. Parker. *Computer Security Management*. Reston Pub. Co., 1981.

[2] Joseph A. Goguen and José Meseguer. Unwinding and inference control. *IEEE Symposium on Security and Privacy*, 0:75, 1984.

[3] William Stallings. *Cryptography and Network Security: Principles and Practice*. Prentice-Hall, 5th edition, 2010.

[4] R. J. Anderson. *Security Engineering: A Guide to Building Dependable Distributed Systems.* Wiley Publishing, 2008.

[5] S. Singh. *The Code Book: the evolution of secrecy from Mary, Queen of Scots, to quantum cryptography.* Fourth Estate, UK, 1999.

[6] Alfred J. Menezes, Paul C. van Oorschot, and Scott A. Vanstone. *Handbook of Applied Cryptography.* CRC Press, 2001.

[7] IACR. `http://www.iacr.org/`.

[8] IEEE Cipher Newsletter, 1994–2011.

2

Security Electronics

E. J. Aas and P. G. Kjeldsberg

Department of Electronics and Telecommunications, NTNU

CONTENTS

2.1   Introduction

Most of the computations needed in the security domain may be done by software. But since we normally search for methods that are hard and/or time-consuming to perform, to avoid intrusion or cracking, it is often beneficial to implement some security applications in hardware. Order(s) of magnitude execution time speedup can be achieved, as demonstrated later in this chapter. High performance is often needed when traffic is high, for example, for authorized money transfers.

Another metric of efficiency is energy consumption. If the system in use is battery operated, we may look for encryption/decryption methods that are parsimonious on energy consumption.

Finally, the cost of the product that performs the encryption/decryption is important. Even though the price of high-performance microprocessors is continuously falling, a miniaturized hardware solution is normally much cheaper, at least when produced in big volumes. The hardware cost is both associated with a nonrecurring development cost, which may be higher than for software, and production cost directly related to the size of the microchip and to the number of chips produced.

There are many reasons why hardware or a combined hardware/software solution may be better than pure software. First of all, software typically runs on a general-purpose microprocessor (implemented in hardware), which is made flexible enough to perform many different tasks. This flexibility comes with a penalty with respect to performance, energy, and component cost. Instruction decoding is necessary to set up the processor to perform the correct task each time instance. In pure hardware, this is not necessary, since this is directly integrated in the application-specific solution. Complex computations that are common in security applications, may in a processor have to be realized as a series of simpler computations which the microprocessor is able to perform. In hardware specifically designed for an application, we can choose to include exactly the complex computation units needed (and only those).

In most cases, an embedded system combining software and dedicated hardware is considered to be most efficient. Before we introduce typical architectures and design flows for security electronics, we will present a few examples of how high performance may be achieved.

A Small Example to Illustrate Computational Complexity

Several methods for encryption/decryption rely on mathematical transformations of digital expressions. The popular RSA encryption/decryption method [1] is one example. In Section 3.4, there is a stringent description of the principles of RSA public key cryptography. In this section, we focus on how we may implement RSA encryption/decryption in efficient ways. Let us take a look at a specific example below.

Encryption is done by computing

$$C = M^e \pmod{n},$$

where
M is the plaintext $\in \{1, 2, \ldots, n-1\}$
n is a product of two large primes, p and q
e is the public key used for encryption
C is the cipher text
    The plaintext M may be retrieved by computing

$$M = C^d \pmod{n},$$

where
d is the private key used for decryption

Notice:

The same computation is performed for encryption and decryption, but with different parameters. The encryption and decryption keys (e, d) are related through the following equation:

$$e * d \pmod{(p-1)(q-1)} = 1.$$

To appreciate the problem of these computations, handling a huge number of digits, we offer a simple example below.

Alice publishes encryption key and modulus thus:$(e; n) = (17; 143)$, but keeps private the value: $d = 113$ (from: $p = 11, q = 13$). She will use the value of d to decrypt incoming messages.

Bob wants to send an encrypted message to Alice. His plaintext message is $M = 50$. The corresponding cipher text is computed thus, using Alice's published values $(e; n)$:

$$C = M^e \pmod{n}$$

$$C = 50^{17} \pmod{143} = 85.$$

Bob sends this message, $C = 85$, to Alice. Now, Alice may decrypt the message from Bob thus:

$$M = C^d \pmod{n}$$
$$M = 85^{113} \pmod{143} = 50.$$

How may we compute such numbers? The naïve way is to compute 85^{113} first; then to perform the modulus operation. Let us study the expression: $M' = 85^{113}$

$M' = 85 * 85 * 85 * 85 * 85 * 85 * 85 * 85 * 85 * 85 * 85 * 85 * 85$
$* 85 * 85 * 85 * 85 * 85 * 85 * 85 * 85 * 85 * 85 * 85 * 85 * 85$
$* 85 * 85 * 85 * 85 * 85 * 85 * 85 * 85 * 85 * 85 * 85 * 85 * 85$
$* 85 * 85 * 85 * 85 * 85 * 85 * 85 * 85 * 85 * 85 * 85 * 85 * 85$
$* 85 * 85 * 85 * 85 * 85 * 85 * 85 * 85 * 85 * 85 * 85 * 85 * 85$
$* 85 * 85 * 85 * 85 * 85 * 85 * 85 * 85 * 85 * 85 * 85 * 85 * 85$
$* 85 * 85 * 85 * 85 * 85 * 85 * 85 * 85 * 85 * 85 * 85 * 85 * 85$
$* 85 * 85 * 85 * 85 * 85 * 85 * 85 * 85 * 85 * 85 * 85 * 85 * 85$
$* 85 * 85 \approx 1.0576418 * 10^{218}$

This is a huge number, and the effort of computing it on a 32 bit or 64 bit computer is significant. Actually, M' consists of more than 700 bits. Remember, our example is small compared to real message handling. Obviously, we must develop more efficient computational methods for hardware implementation. We get help from certain mathematical relations for modulus computation. The most important trick is to reduce the number of bits in processing whenever possible. One important relation is

$$(A * B) \pmod{n} = (A \pmod{n}) * (B \pmod{n}).$$

Obviously, we may rewrite the expression for M' above. One idea is to exploit squaring. For example, we may write
$85^{64} = ((((((85^2)^2)^2)^2)^2)^2),$
and perform $\pmod{n}$ whenever possible, leading to a reduction of the number of bits to handle at any time. In addition to attaining efficient computations, the modulo operations prevent overflow of computation.

2.2 Examples of Security Electronics

2.2.1 RSA as Hardwired Electronics

Given the example of RSA encryption in Section 2.1, we are now ready to study some efficient implementations of the equations for C and M. Reference [2] is used as an inspiration.

Algorithm: **RL binary method**
Input: $M; e; n$
Output: $C := M^e \pmod{n}$
1. $C := 1;$ $P := M$
2. for $i = 0$ to $h - 2$
2a.   if $e_i = 1$ then $C := C * P \pmod{n}$
2b. $P := P * P \pmod{n}$
3. if $e_{h-1} = 1$ then $C := C * P \pmod{n}$
4. return C

FIGURE 2.1
The RL binary method.

TABLE 2.1
Execution of the RL binary method

i	e_i	Step 2a (C)	Step 2b (P)
0	1	$1 * M = M$	$(M)^2 = M^2$
1	1	$M * M^2 = M^3$	$(M^2)^2 = M^4$
2	0	M^3	$(M^4)^2 = M^8$

Step 3: $e_{h-1} = 1$, hence $C = (M^3) * (M^8) = M^{11}$

One possibility for the computation is to use the RL binary method (RL: right to left) as given in Figure 2.1.

In the RL algorithm, we are scanning e from its least significant bit ($i = 0$), to its most significant bit ($i = h - 1$). Modular squaring is performed for each bit. Modular multiplication is done whenever the bit is 1.

To demonstrate how the RL algorithms works, let us use $h = 4$, $e = 11$ ($= 1011B$), thus computing M^{11}. The RL algorithm starts with $C := 1$ and $P := M$ and proceeds as in Table 2.1. We end with the expected value, M^{11}. Notice that in a real implementation, we will perform the modulo operation at each step.

The next task is to consider how the RL algorithm may be implemented as efficiently as possible. Our primary concern is time spent for the execution, which will tell us how many messages we may encrypt/decrypt per second. Note that we need three registers: M, C, and P to store variable values (we may let P and M share one register, if M is not needed thereafter).

The total time spent depends on the following

- We perform $(h - 1)$ squarings, and up to $(h - 1)$ multiplications; on the average $(h - 1)/2$ multiplications.

- In addition, since Step 2a and Step 2b are performed independently of each other, they may be parallelized. By using two multipliers (or one multiplier and one squarer), we realize that the total time is constrained by computing $(h - 1)$ squaring operations on k-bit integers.

In fact, there exist advanced algorithms for computing modular exponentiation, see [2], which are slightly faster than the RL binary method. But for our purpose, it suffices to consider the RL binary method only.

What kind of arithmetic modules do we need to execute the RL binary method?

We need: modular multipliers, and modular multipliers with reduction. The reduction of integers $> n$ is done by subtraction or addition.

The design space, that is, the realm of feasible designs to realize the RL binary method, is huge. We may select different architectures of arithmetic modules, depending on whether we want high performance or low silicon area, or low energy consumption. Normally, the trade-offs between these are trade-offs between cost (in terms of silicon area) and speed.

For modular multiplication, we may exploit multiply and divide, or alternate multiplication and reduction, Brickell's method, Montgomery's method, or others. Adders may be of many kinds: carry propagate (compact and slow), carry look-ahead (more area, but faster), carry save, carry select, and several more. In [2], the reader may find several efficient ways of performing modular exponentiation and multiplication. We will not go through all these, but give a specific example of realization below. The example will provide insight into performance and cost as a function of complexity driven by key length.

From a project in the course *Realization and Test of Digital Components*, NTNU, Fall 2008, we retrieved the following insight from one of the best project reports, written by Rune A. Bjørnerud and Lars O. Opkvitne. The main focus of the specification was maximum throughput, while constraining the use of resources on an FPGA. The realization was done on an FPGA from Xilinx: the Virtex 2 XC2V1000 [3]. The main resources available are lookup tables (LUTs), and registers.

The throughput, measured as the number of messages encrypted/decrypted per second, is computed by the expression:

$$Tpm = f/Ncl \qquad [\#msgs/s],$$

where f is the clock frequency and Ncl is the total number of clock cycles needed to encrypt or decrypt the message.

A more common metric of throughput is defined as the number of *bits* encrypted/decrypted per second. This is computed by the expression:

$$Tpb = m * f/Ncl \qquad [b/s],$$

where m is the length of the message in bits. For the optimized design, Ncl and h are related through

$$Ncl = h^2 + 8h + 24.$$

We observe that for realistic key sizes h, Ncl increases quadratically with h. What about the clock frequency? Obviously, it will decrease as h increases, since there is more hardware involved. It is not feasible to reach an analytical

```
Device utilization summary:
------------------------------

Selected Device : 2v1000fg456-4

 Number of Slices:                    3041  out of    5120    59%
 Number of Slice Flip Flops:          2568  out of   10240    25%
 Number of 4 input LUTs:              4928  out of   10240    48%
 Number of IOs:                         69
 Number of bonded IOBs:                 69  out of     324    21%
 Number of GCLKs:                        1  out of      16     6%
```

FIGURE 2.2

Excerpts of a Log file generated by the synthesizer Xilinx ISE. Key length $h = 128$ bit. No. of LUTs used is encircled.

TABLE 2.2

Results from synthesis with Xilinx ISE 9.2

No. of bits h	Frequency [MHz]	No. of clock cycles	Throughput [msgs/s]	Throughput [kbits/s]
128	61.6	17449	3523	451
256	42.7	67380	1222	313
512	26.7	265807	382	196
1024	15.4	1055957	110	116

expression for this relation. The limiting factor for maximum frequency is the so-called critical path delay. This is a path between registers clocked by the same clock (we assume only one system clock generator is this case). The critical path typically goes through the adder system.

We employ synthesis programs that translate VHDL code of the system into FPGA code. By selecting proper optimization parameters, one will attain a very efficient hardware version, with high throughput and constrained use of resources.

We present results for various values of h below. First, Figure 2.2 presents a device utilization summary, showing how much resources a given implementation will use.

A complexity analysis of frequency, area and throughput was performed for the block sizes $128, 256, 512,$ and 1024 (the device utilization summary for block sizes larger than 128 is not shown). The synthesis into FPGA realization was done with the synthesis program Xilinx ISE 9.2 [3]. Table 2.2 presents the impact of the number of bits on frequency, number of clock cycles, and the resulting throughput when the message is of the same length as the key; $m = h$.

Note that if we realize the circuit as a custom-designed integrated circuit with state-of-the-art technology, we will achieve higher performance than on the FPGA. But the general trend showed in Table 2.2 will still be valid.

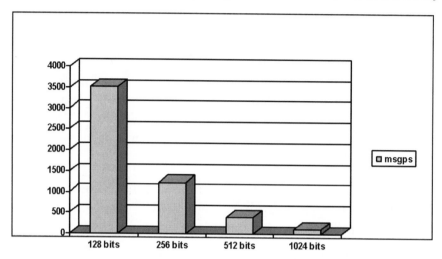

FIGURE 2.3
Throughput [messages/second] as a function of message and key length.

In Figure 2.3, we have shown the throughput from Table 2.2 as a chart. We observe that the throughput, in terms of number of messages/second, decreases dramatically as key and message length increase.

2.2.2   AES as Hardwired Electronics

The Advanced Encryption Standard (AES) is a crypto-standard defined by the National Institute for Standards and Technology (NIST) in the United States. It is a result of a contest organized in 1997. Three years later, the Rijndael algorithm, designed by the two Belgians, Joan Daemen and Vincent Rijmen, was announced as winner [4] [5]. The final AES standard was presented in 2001 [6]. With a few small modifications, it is based on the Rijndael algorithm. An important reason for its success was the fact that it is relatively easy to implement both in software and hardware.

While the RSA algorithm described in the previous section is an asymmetric cipher with different keys for encryption and decryption, AES is a symmetric cipher where the same key is used on both sides. It has a fixed message block size of 128 bits of text (plain or cipher), and keys of length 128, 192, or 256 bits. When longer messages are sent, they are divided into 128 bit blocks. Obviously, longer keys make the cipher more difficult to break, but also enforce a longer encrypt and decrypt process.

A 128 bit message block can be thought of as organized in a two-dimensional 4∗4 byte array. AES encryption has four main operations that are repeated multiple times, and work on bytes, rows, and columns of this array: AddRoundKey, SubBytes (substitute bytes), ShiftRows, and MixColumns.

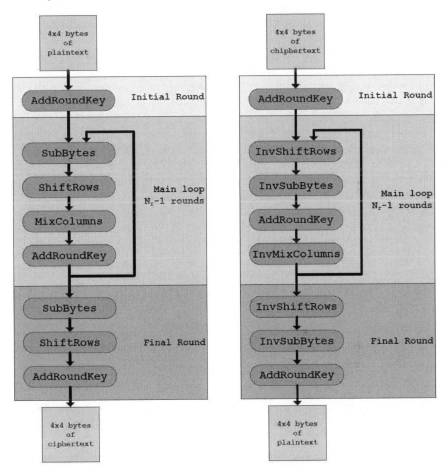

FIGURE 2.4
AES encryption and decryption [7].

AES decryption also encompasses four main operations, similar to encryption. These are AddRoundKey, InvSubBytes, InvShiftRows, and InvMixColumns. The number of times each operation is performed depends on the key length, as depicted in Figure 2.4. N_r is 10, 12, and 14 for key lengths 128, 192, and 256, respectively.

We will now take a look at some of the steps in the algorithm to see what makes them suitable for hardware implementation. A number of students at NTNU have worked on implementation of AES as their master's thesis projects [7][8][9]. Several of the examples and results will be taken from them. We will not give a complete description of each operation. For interested readers there is also a very large number of alternative implementations and implementation methodologies for AES to be found in the literature, for example, [10], [11] [12] [13] [14].

The AddRoundKey operation performs an XOR operation between the message block and a separate 128 bit round key for each of the N_r rounds (plus one as can be seen in Figure 2.4). These round keys are generated from the original 128, 192, or 256 bit keys (actually from encoder key $N_r + 1$ for decoding). It is possible to generate all keys in advance and store them in a memory of up to 240 bytes. This may sound small, but in some application areas, like smart cards, it can still be too much. Accessing memory can also be both time and energy consuming. With an efficient on-the-fly key expansion in hardware it can therefore, for example, be better to generate the next round key in parallel with execution of the current encryption round [7]. This is impossible in software, since a standard processor can only perform one operation at a time. Furthermore, even though simple to implement in hardware, the many XOR operations in AddRoundKey can take a lot of time in software. In the most extreme case, the processor can only perform a one bit XOR at a time, yielding 128 clock cycles per AddRoundKey operation. More typically the processor can handle 8, 16, or 32 bits at a time, but this will still require from 4 to 16 clock cycles. A fully parallel hardware implementation can perform XOR on all 128 bits in one clock cycle, or on a number of bits adapted to the rest of the application specific architecture.

Both the SubByte and InvSubBye operation and the Key Expansion use a so-called Rijndael S-box. The S-box substitutes a one byte state (that is, an eight bit binary number) with another state (another eight bit number). This can easily be done using two 256 byte look-up-tables (LUT), one for SubByte and one for InvSubByte. The new state for each possible current state is stored in the LUT. The current state is then used to address the correct location in the LUT holding the new state. As with the key storage, this 512 byte memory can be prohibitively large or may consume too much power when accessed intensively. Alternatively, one can exploit the fact that the new state values can be calculated using affine transforms and multiplicative inversion using isomorphic mapping [14]. Though mathematically complex, they can be implemented in hardware with low resource usage, both with respect to area and energy. Reference [7] reports results from three experiments. The first uses two LUTs. The second uses one LUT for the multiplicative inversion, which is the same for both SubBytes and InvSubBytes. In addition, it implements affine transformations directly to reach the correct end result. The third experiment does not use any LUT at all, instead implementing all the mathematical calculations directly. Table 2.3 gives the results for the three experiments with area reported as number of two-input NAND-gate equivalents, power in μW, and delay through the circuits in ns. Synthesis was done in a typical 90 nm circuit technology. As can be seen, there is a trade-off between area and power on one hand and delay on the other. As long as the target frequency of the system is reached with the version without any LUTs, this version is preferred.

We have now seen some examples of how elements of the AES algorithm

TABLE 2.3

Comparison of Sbox implementations [7]

	Area (NAND2 eq.)	Power [μW]	Delay [ns]
2 LUT	1408	95.49	1.15
1 LUT	821	88.43	2.30
No LUT	300	64.19	2.93

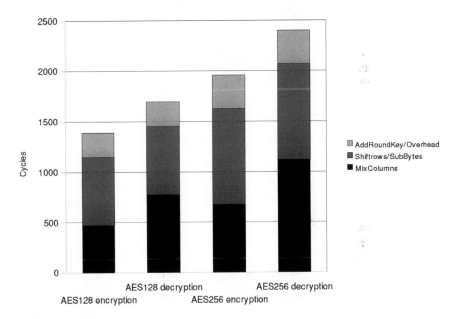

FIGURE 2.5

Cycle-count for AES encryption and decryption implemented in software [7].

can be implemented in hardware. This is not meant to be a complete coverage of all aspects regarding hardware design, but should illustrate some of the considerations we need to take into account. Finally, we will now compare complete implementations of the algorithm to see examples of what can be achieved here.

In [7], a pure software and a combined hardware/software implementation of AES is presented. The software version is based on the work presented in [11] with some additional improvements. Figure 2.5 shows the number of cycles needed to encrypt and decrypt a 128 bit data block for this version when run on a 32 bit ARM Coretex M3 processor [15].

Similarly, the hardware part of the combined hardware/software solution is based on [14] with a number of additional improvements, including extension to 256-bit keys. The datapath is 32 bit wide, with the ability of performing (Inv)MixColumn, AddRoundKey, and (Inv)SubBytes in parallel in one cycle. Four cycles are needed for each round of a 128 bit data block. One additional

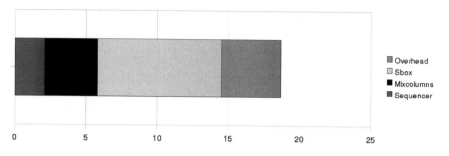

FIGURE 2.6
Delay in *ns* through hardware datapath [7].

cycle is used to perform (Inv)ShiftRows. For complete encryption or decryption of a 128 bit data block, 55 cycles are required in AES 128 mode, and 75 cycles in AES 256 mode. Figure 2.6 shows the delay in ns through the hardware datapath when synthesized using an ARM Sage-X 180 nm standard cell library. It allows a maximal clock frequency of 43.5 MHz. The datapath is assumed controlled by a microcontroller, which makes blocks of data and the corresponding keys available whenever needed. In the meantime, it can perform other useful activities or be in an idle low power mode.

To compare the two implementations, pure software and combined hardware/software, a clock frequency of 32 MHz is assumed for throughput calculations. For energy consumption in software, the number of cycles needed is multiplied with the average energy per cycle, as given by the data sheet for the ARM Coretex M3 processor when synthesized with the ARM Sage-X 180 nm standard cell library [15]. For hardware, energy consumption calculations is based on simulation using Mentor Graphic's Modelsim [16] and Synopsys' Power Compiler [17] with the same standard cell library. Energy consumption on buses and in memories are assumed to be similar for both implementations and are hence ignored for the comparison.

Table 2.4 reports the results from the experiments, showing a very large percentage reduction in energy consumption, and increase in throughput when going from a software solution to a hardware solution. The numbers here may not be fully exact, since they are based on data sheets and simulations only, but still they give a good indication of what is achievable. Obviously, if energy is not an issue, and you already have a microprocessor in your system, you should use the software solution if you do not need a higher throughput than what it can deliver. If your system is battery operated, or you need a higher throughput, you will definitely need to go for hardware.

TABLE 2.4

Comparison of software and software/hardware solution [7]

	Software	Hardware	Percentage
AES 128 encryption, [nJ/128bits]	333	8.0	2.4 %
AES 128 decryption, [nJ/128bits]	407	9.1	2.2 %
AES 256 encryption, [nJ/128bits]	469	11.0	2.3 %
AES 256 decryption, [nJ/128bits]	576	12.8	2.2 %
Throughput AES 128 encryption [Mbps]	3.0	74.5	2483 %
Throughput AES 128 decryption [Mbps]	2.4	74.5	3104 %
Throughput AES 256 encryption [Mbps]	2.1	54.6	2600 %
Throughput AES 256 decryption [Mbps]	1.7	54.6	3212 %

2.2.3 Examples of Commercial Applications

Disk Encryption of Hard Disk Drives

Portable devices like laptops and desktop computers, may be stolen or invaded electronically. An intruder may then get access to sensitive data on the hard disk. But if the data entering the disk is always encrypted, then it may be securely safeguarded. There exist commercial products to encrypt and decrypt at the user's will. One solution offers a system to protect the hard disk this way. A Key Management System based on, for example, a USB pin for key entry provides a user-friendly encryption technology. A 256 bit key system within the AES algorithm is provided; see [18] for details.

Crypto-authentication

Another product group is so-called crypto-authentication. There are several applications where validation of physical or logical elements is important. Here are some examples based on hashing algorithms [19]:

- Prevent cloning of consumables such as battery packs, filters, test strips, ink cartridges.

- Store an unmodifiable serial number; provide a verification method to identify fraudulent copies.

- Validate the integrity and authenticity of firmware, software, or other data such as media, including download protection.

- Securely exchange session keys for encrypted communications over wireless or wired networks.

- High security logical or physical access control, replacement, or augmentation to passwords.

2.3   Side Channel Attacks

A side channel attack involves gathering information regarding a crypto-system, its keys, or its secret messages, through monitoring of other system artifacts than the encrypted message itself. The information gathered through these channels can assist in breaking a code even if it does not give enough information on its own.

When a crypto-algorithm is implemented in hardware, or in software run on a hardware microprocessor, it uses power and emits electromagnetic radiation while encrypting and decrypting a message. This opens for power analysis attacks [20] [21] [22] or even attacks based on "seeing "the screen picture at a distance using directional antennas that detects radiation [23]. In some cases, the time encryption and decryption take can be data dependent, which can make them vulnerable for timing attacks [24].

While designing hardware that performs encryption and decryption, it is important to take the possibility of side channel attacks into consideration. The system can be made less vulnerable by making sure that all operations take the same amount of time and consumes the same amount of power independently of the data that is being processed. An example of a coprocessor for AES (see Section 2.2.2) immune to first-order differential power analysis is given in [25]. Additional details are outside the scope of this book but can also be found in other references in this chapter.

2.4   Summary

In this chapter, we have focused on hardware realizations of security electronics. In particular, we have studied encryption/decryption methods, and shown examples of implementations. RSA and AES methods have been used for illustration. We have emphasized the benefits of hardware over software with respect to performance and power consumption. Comparisons between software solutions, hardware solutions, and mixed HW/SW implementations have been offered.

If energy is not an issue and there is a microprocessor in your system, you should use the software solution, if you do not need higher throughput than it can master. If your system is battery operated, or you need a higher throughput, you will definitely need to go for hardware. We have shown examples of almost 50 times energy reduction, and 30 times throughput improvement.

2.5 Further Reading and Web Sites

Much research is being performed in the field of security electronics. Important parts of this is reported at the annual Workshop on Cryptographic Hardware and Embedded Systems (http://www.iacr.org/workshops/ches/) with proceedings published by Springer in its Lecture Notes in Computer Science.

For information regarding cryptography implementations in specific application domains, the books *Embedded Security in Cars* [26] and *Smart Cards, Tokens, Security and Applications* [27] give good examples.

The book *Embedded Cryptographic Hardware: Design and Security* [28] is by now a few years old but covers many aspects related to the content of this chapter.

Finally, Wikipedia (http://www.wikipedia.org/) has a number of interesting articles related to cryptography in general, and algorithms such as AES, RSA, and DES, and their implementation, in particular.

Bibliography

[1] C. K. Koc. RSA hardware implementation. Technical Report TR 801, RSA Laboratories, Redwood City, California, USA, April 1996. ftp://ftp.rsasecurity.com/pub/pdfs/tr801.pdf.

[2] C. K. Koc. High-speed RSA implementation. Technical Report TR 201, RSA Laboratories, Redwood City, California, USA, November 1994. ftp://ftp.rsasecurity.com/pub/pdfs/tr201.pdf.

[3] Xilinx web page. http://www.xilinx.com/, March 2010.

[4] J. Daemen and V. Rijmen. The block cipher Rijndael. In *Proceedings of the Fourth Working Conference on Smart Card Research and Advanced Applications*, volume 1820 of *Lecture Notes in Computer Science*, pages 277–284, Bristol, United Kingdom, September 2000. Springer Verlag.

[5] J. Daemen and V. Rijmen. *The Design of Rijndael—AES—The Advanced Encryption Standard*. Springer, Berlin Heidelberg, Germany, 2002.

[6] NIST. Announcing the ADVANCED ENCRYPTION STANDARD (AES). Technical Report FIPS-197, National Institute of Standards and Technology, USA, November 2001. http://csrc.nist.gov/publications/fips/fips197/fips-197.pdf.

[7] Ø. Ekelund. Low energy AES hardware for microcontroller. Master's thesis, Department of Electronics and Telecommunications, Norwegian

University of Science and Technology, Trondheim, Norway, July 2009. http://daim.idi.ntnu.no/masteroppgave?id=4658.

[8] R. C. Larsen. Low cost hardware acceleration of cryptographic algorithms in microcontroller. Master's thesis, Department of Electronics and Telecommunications, Norwegian University of Science and Technology, Trondheim, Norway, June 2007.

[9] T. B. Alu. VHDL-implementation of advanced encryption standard (AES). Master's thesis, Department of Electronics and Telecommunications, Norwegian University of Science and Technology, Trondheim, Norway, June 2005.

[10] S. Ahuja, S. T. Gurumani, C. Spackman, and S. K. Shukla. Hardware coprocessor synthesis from an ANSI C specification. *IEEE Design and Test of Computers*, 26(4):58–67, July 2009.

[11] G. Bertoni, L. Breveglieri, P. Fragneto, M. Macchetti, and S. Marchesin. Efficient software implementation of AES on 32-bit platforms. In *Revised Papers from the 4th Workshop on Cryptographic Hardware and Embedded Systems*, volume 2523 of *Lecture notes in computer science*, pages 159–171, Redwood City, California, USA, August 2002. Springer Verlag.

[12] S. Mangard, M. Aigner, and S. Dominikus. A highly regular and scalable aes hardware architecture. *IEEE Trans. on Computers*, 52(4):483–491, April 2003.

[13] S. O'Melia and A. J. Elbirt. Enhancing the performance of symmetric-key cryptography via instruction set extensions. *To be published in IEEE Trans. on Very Large Scale Integration (VLSI) Systems*, PP(99):1–14.

[14] A. Satoh, S. Morioka, K. Takano, and S. Munetoh. A compact Rijndael hardware architecture with s-box optimization. In *Proceedings of the 7th International Conference on the Theory and Application of Cryptology and Information Security*, volume 2248 of *Lecture Notes in Computer Science*, pages 239–254, Gold Coast, Australia, December 2001. Springer Verlag.

[15] ARM. *An Introduction to the ARM Cortex-M3 Processor*, October 2006. www.arm.com/files/pdf/IntroToCortex-M3.pdf.

[16] Mentor grapics web page. http://www.mentor.com/, March 2010.

[17] Synopsys web page. http://www.synopsys.com/, March 2010.

[18] HDD. High density devices web page. http://www.hdd.no/, March 2010.

[19] Atmel. Web page on secure authentication. `http://www.atmel.com/products/cryptoauthentication/`, March 2010.

[20] P. Kocher, J. Jaffe, and B. Jun. Differential power analysis. In *Proceedings of the 19th Annual International Cryptology Conference on Advances in Cryptology*, volume 1666 of *Lecture Notes in Computer Science*, pages 388–397, Santa Barbara, Calif, USA, August 1999. Springer Verlag.

[21] L. Goubin and J. Patarin. DES and differential power analysis (the duplication method). In *Proceedings of the First International Workshop on Cryptographic Hardware and Embedded Systems*, volume 1717 of *Lecture Notes in Computer Science*, pages 158–172, Worcester, Massachusetts, USA, August 1999. Springer Verlag.

[22] T. S. Messerges, E. A. Dabbish, and R. H. Sloan. Examining smart-card security under the threat of power analysis attacks. *IEEE Trans. on Computers*, 51(5):541–552, May 2002.

[23] M. G. Kuhn. Electromagnetic eavesdropping risks of flat-panel displays. In *Proceedings of the 4th Workshop on Privacy Enhancing Technologies*, volume 3424 of *Lecture Notes in Computer Science*, pages 23–25, Toronto, canada, May 2004. Springer Verlag.

[24] P. Kocher. Timing attacks on implementations of Diffie-Hellman, RSA, DSS, and other systems. In *Proceedings of the 16th Annual International Cryptology Conference on Advances in Cryptology*, volume 1109 of *Lecture Notes in Computer Science*, pages 104–113, Santa Barbara, Calif, USA, August 1996. Springer Verlag.

[25] E. Trichina, T. Korkishko, and K. H. Lee. Small size, low power, side channel-immune AES coprocessor: Design and synthesis results. In *Revised Selected and Invited Papers from the 4th International Conference, AES 2004*, volume 3373 of *Lecture Notes in Computer Science*, pages 113–127, Bonn, Germany, May 2004. Springer Verlag.

[26] K. Lemke, C. Paar, and M. Wolf (Eds.). *Embedded Security in Cars– Securing Current and Future Automotive IT Applications*. Springer, Berlin Heidelberg, Germany, 2006.

[27] K. Mayes and K. Markantonakis (Eds.). *Smart Cards, Tokens, Security and Applications*. Springer, Berlin Heidelberg, Germany, 2008.

[28] N. Nedjah and L. de Macedo Mourelle (Eds.). *Embedded Cryptographic Hardware: Design and Security*. Nova Science Publishers, New York, USA, 2005.

3

Public Key Cryptography

S. O. Smalø

Department of Mathematical Sciences, NTNU

CONTENTS

In this chapter, the notion of public key cryptography is introduced. Then the RSA-scheme introduced by Rivest, Shamir, and Adleman is treated. The security of this scheme is based on factorization problems of integers. We also include illustrations of how another number theoretical problem, known as the discrete logarithm problem, is used and implemented in the public key cryptographic systems today.

3.1 Introduction

Let us look at three persons, named A, B, and E, who can communicate over an open communication channel. How should we organize this communication such that each person can read the messages intended for him, but get no clue about the content of a messages he intentionally or accidentally get in his possession and which are intended for someone else? In addition, we want to set up the system such that a person getting a message intended for him should also be sure who is the original sender. The third property we would like this system to have is that the sender later cannot refuse that he is the one who originally sent the message.

In this problem, there are three layers of security. The first layer is an encryption ensuring that nobody except the intended person can read the message, the second layer is that the intended reader can verify who the sender of the message is, and the third layer is the signature of the sender making it impossible for him to deny that he is the real origin of the message.

In former days, one could achieve this by sending an entrusted person with an envelope with seal-wax and seal, and including the signature of the sender on the message. In the military, where there might be a chance that the enemy could catch the messenger, the content was distorted in a prescribed way so the enemy could not read the message. It was then important that the messenger did not know the content of the message nor knowing how to restore it into a readable form. Therefore, the intended receiver and the sender had to agree beforehand how to do the manipulation both for encryption and decryption.

A way of doing this that goes back to the Roman emperor Caesar was just to shift the alphabet three places, so instead of writing an A, it was written a D and instead of a B, an E was written, and then repeating this for the other letters all the way until one wraps the alphabet around so that an X was substituted with an A, a Y was substituted with a B, and a Z was substituted with a C. The receiver could then substitute back so that when he saw the signature FDHVDU, he could reinterpret this by shifting back three letterers getting CAESAR. However, this system has almost no security. Modifications of this system have been tried, but the structure in a natural language makes it hard to get adequate security.

To get started, we may as well change our messages written in a natural language like English or Norwegian into a form better suited for mathematical manipulation.

This may be obtained by substituting the symbols on a keyboard by numbers. One can do this for example using ASCII, where each symbol on the keyboard is substituted with an 8-digit binary number.

We can now do this transition and then break this down into blocks of a given size, and fill in with zeros to obtain the prescribed length of each block. For example, a sentence is broken down into sequences consisting of 100 symbols, including letters, space, brackets, and punctuations, and then each block will be represented as a 800-digit number in binary form.

In this way, we can come from a message in a complicated language to the simple nature of numbers given in binary form. We are now going to take a detour into digital representation before returning to public key cryptography.

3.2   Hash Functions and One Time Pads.

So we can now look at a sequence of 800-digit numbers as the message which A want to send to B. If A and B has had a previous private rendezvous they could have exchanged in advance an 800-digit binary number, $h(A, B)$, which they could use for safe communication. This 800-digit number $h(A, B)$ is called the "secret key" for A and B. When A wants to send the message consisting of the sequence $a_1, a_2, \cdots, a_m$ of m 800 digit numbers a_i to B, he instead sends the sequence

$$a_1 + h(A, B), a_2 + h(A, B), \cdots, a_m + h(A, B).$$

Here the addition is done modulo 2, meaning that he follows the following rules digit by digit $0 + 0 = 0$, $1 + 0 = 0 + 1 = 1$, and $1 + 1 = 0$. When B gets the message consisting of the sequence $a_1 + h(A, B), a_2 + h(A, B), \cdots, a_m + h(A, B)$

of m 800-digit numbers, and she suspects that it is coming from A, she adds the same number $h(A, B)$, which she has in her possession to each of the blocks in the sequence and the sequence

$$a_1 + h(A, B) + h(A, B), \cdots, a_m + h(A, B) + h(A, B) = a_1, \cdots, a_m$$

will appeared in front of B. This was the original message A wanted to transfer to B.

This has a couple of drawbacks, and the first one is that A and B have to meet in order to exchange the "secret key" $h(A, B)$, and they have to store it in a safe place where nobody can get hold of it. Also, if one person knows one block a and how it is transformed $a + h(A, B)$, he knows the "secret key" $h(A, B)$ of A and B. This follows since $a + a + h(A, B) = h(A, B)$. Besides, if A is going to communicate with many persons this way, the administration of all his secret keys will be a problem. There is one more weakness of this system, and that is that if the "secret key" $h(A, B)$ is used repeatedly, then a third person will get some information about the key each time it is used if there is any structure in the messages, like it is in a natural language.

However, if such a "key" is used only once to encrypt only one block, it is totally safe, and therefore it is also used that way. It is then called *a one time pad*. Such one time pads were used in diplomatic and military correspondence where absolute security was needed.

A partial solution to the problem related to reuse of the secret key is to create an arbitrary, long, seemingly random key from a secret key two persons are sharing. In this connection, a special type of function called hash functions will be introduced.

A hash function is a function h from the natural numbers $\mathbb{N}$ to a given interval $\{i \in \mathbb{N} \mid 0 \leq i \leq N\}$, and which are supposed to satisfy the following three condition:

1) It should be easy and fast to calculate the number $h(x)$ from the number x.

2) To each number y in $\{i \in \mathbb{N} \mid 0 \leq i \leq N\}$ it should be hard to find a solution x to the equation $f(x) = y$, that is, it should be hard to find an x mapping onto y by f.

3) It should be hard to find two numbers x and z such that $f(x) = f(z)$.

Clearly, if property 2) is not valid, then property 3) is not valid either. Hence, it is the property 3) which is the crucial one, and which is also hardest to obtain in practice.

One immediately sees that the number N have to be rather large, for otherwise one can just calculate $f(x)$ for enough numbers until one get the same number twice, which then would violate property 3). Many proposed hash functions and also many functions which have been in use as such functions do not satisfy property 3). Of course, a hash function in the 19th century could be rather primitive since the calculation power was so low, but remember, it should still be relatively fast to calculate $f(x)$ when x is given.

Before we go on, we need some modular arithmetic. For a positive number

n, we consider the integers in the interval from 0 to $n-1$. From the Euclidean division algorithm, we know that we can write any number m as $m = qn + r$ with r, the reminder, satisfying the inequalities $0 \le r \le n-1$. Moreover, q and r are uniquely determined by m when this additional property $0 \le r \le n-1$ is satisfied. We now define a function from the integers to the interval $\{r \mid 0 \le r \le n-1\}$ by sending any integer m to the number r in the above equation. We will from now on denote this number r by $m \bmod n$. We introduce the notation $\mathbb{Z}_n$ for the set of integers in the interval from 0 to $n-1$, that is, $\mathbb{Z}_n = \{r \in \mathbb{Z} \mid 0 \le r \le n-1\}$. Among the numbers in $\mathbb{Z}_n$ we introduce an operation which we in the beginning will denote by $+_n$ and call addition modulo n. Here $x +_n y = x + y$ if $x + y \le n-1$ and $x +_n y = x + y - n$ if $x + y \ge n$. In other words, we let $x +_n y = (x + y) \bmod n$ be the remainder one get by dividing $x + y$ by n.

In the same way one can define the operation $\cdot_n$ by letting $x \cdot_n y = (x \cdot y) \bmod n$ be the remainder one gets by division of $x \cdot y$ by n.

For example, for $n = 11$, we have that $5 +_{11} 8 = 2$ since $5 + 8 = 13 = 1 \cdot 11 + 2$, and $5 \cdot_{11} 7 = 2$ since $5 \cdot 7 = 35 = 3 \cdot 11 + 2$.

We also introduce the operation of taking powers with respect to n for a positive integer m and an x in $\mathbb{Z}_n$ by letting (x^m) be the result $x \cdot_n x \cdot_n \cdots \cdot_n x$, where there are m factors all equal to x. By doing this, one ends up with a system where most of the usual arithmetic operations for integers hold. We list the rules that are generally true in this setting and give some few examples of some rules that hold for integers, but which are not always true in the situation of $\mathbb{Z}_n$ with the operations $+_n$ and $\cdot_n$.

1) $(x +_n y) +_n z = x +_n (y +_n z)$ holds for all x, y and z in $\mathbb{Z}_n$.

2) $0 +_n x = x +_n 0 = x$ for all x in $\mathbb{Z}_n$.

3) $x +_n (n - x) \bmod n = 0$ for all x in $\mathbb{Z}_n$.

4) $x +_n y = y +_n x$ for all x and y in $\mathbb{Z}_n$.

That these four properties are satisfied is usually expressed by the term that $(\mathbb{Z}_n, +_n)$ is an abelian group under addition modulo n.

5) $(x \cdot_n y) \cdot_n z = x \cdot_n (y \cdot_n z)$ for all x, y and z in $\mathbb{Z}_n$.

6) $1 \cdot_n x = x$ for all x in $\mathbb{Z}_n$.

7) $x \cdot_n (y +_n z) = (x \cdot_n y) +_n (x \cdot_n z)$ and $(x +_n y) \cdot_n z = (x \cdot_n y) +_n (x \cdot_n z)$ for all x, y and z in $\mathbb{Z}_n$.

8) $x \cdot_n y = y \cdot_n x$ for all x and y in $\mathbb{Z}_n$.

A system satisfying all these eight properties is usually called a commutative ring. If the last property is not satisfied, one just calls such a system a ring.

Here we might have that $x \cdot_n y = 0$, where both $x \ne 0$ and $y \ne 0$. For example, $2 \cdot_6 3 = 0$, and $2 \cdot_4 2 = 0$ without any of the factors being 0.

Two integers x and y are said to be relative prime if the greatest common divisor of x and y is 1.

A result of the 17th-century French mathematician Fermat states that if p is a prime number then $x^p = x$ in $\mathbb{Z}_p$ for all x in $\mathbb{Z}_p$. This was later generalized by the 18th-century Swiss mathematician Euler who proved that for an integer

x relative prime to n in $\mathbb{Z}_n$, $(x^{\phi(n)}) = 1$ in $\mathbb{Z}_n$, where $\phi(n)$ is the number of integers in $\mathbb{Z}_n$ which are relative prime to n. This can be extended to the following result using the Chinese remainder theorem which was known by Sun Tzu in China 1700 years ago: If n is a positive integer not divisible by any square number except 1, then the equality $x^{z \cdot \phi(n)+1} = x$ holds for each x in $\mathbb{Z}_n$ and each integer z.

Example

The function f which associates to each number m the remainder of m by division by n satisfies the property 1) which we want a hash function to satisfy, but none of the other properties we want a hash function to satisfy. One sees this immediately since $f(r) = r = f(n+r)$ for all $0 \le r \le n - 1$.

However, there are some indications that one can still use some of the operations from $\mathbb{Z}_p$ when p is a large prime number to create some hash functions. This is based on some number theoretical facts which we will state in the same way as the result of Fermat and Euler we have included without a proof. To a given prime number p, there always exists a number $\alpha \in \mathbb{Z}_p$ such that the set of numbers $\{(\alpha^i) \in \mathbb{Z}_p \mid 0 \le i \le p - 2\}$ is equal to the set $\{j \mid 1 \le j \le p-1\}$. Therefore, the function g given by composing the function $f : \mathbb{Z} \to \mathbb{Z}_{p-1}$ by taking the remainder modulo $p - 1$ with the exponential function with base α, giving $g(i) = (\alpha^{f(i)}) \bmod p$ gives a function from the nonzero integers onto the nonzero elements of $\mathbb{Z}_p$. This function is fast to calculate as we will soon indicate. However, there seems to be no effective way of expressing the inverse of the exponential function, that is, finding a solution of the equation $y = \alpha^x$ with y a known entity in $\mathbb{Z}_p$. However, the third property of a hash function is not satisfied since two natural numbers where the difference is divisible by $p - 1$ gives the same result. The problem of finding the inverse of the function given by sending x in $\mathbb{Z}_p$ to α^x is known as the discrete logarithm problem. An integer α with the property that the set $\{\alpha^i \bmod p \mid 0 \le i \le p - 1\}$ is equal to $\{r \mid r \in \mathbb{Z}_p \setminus 0\}$ is called a primitive element modulo p, or a $(p - 1)$th primitive root of unity modulo p.

Example: Consider the number 7. Then $2, 2^2 = 4, 2^3 = 1, 2^4 = 2$ so we get that $\{2^i \mid i = 1, 2, 3, 4, 5, 6\} = \{1, 2, 4\}$ in $\mathbb{Z}_7$, so 2 is not a primitive root modulo 7. If we instead take the number 3, we get $3, 3^2 = 2, 3^3 = 6, 3^4 = 4, 3^5 = 5, 3^6 = 1$, which shows that 3 is a primitive root modulo 7.

The number of primitive roots modulo a prime number p is the same as the number of numbers in $\mathbb{Z}_{p-1}$ relative prime to $p - 1$. This is expressed by the Euler ϕ-function that we introduced above, evaluated at $p-1$ as $\phi(p-1)$. This number is less than or equal to $(p-1)/2$ when p is an odd prime number since $p - 1$ is then an even number. This discrete logarithm problem can be used to produce a "hash"-function from a bounded interval onto an interval where the number of decimals is approximately cut in half.

This can be done the following way: Find prime numbers p and q such that $p = 2q + 1$ with, say, 100 digits. Then the hash function we will look at takes as input numbers with less than 200 digits and produces a number with less than 100 digits.

Choose two primitive roots α and β modulo p. Then $(\alpha^a) = \beta$ for an a which cannot be calculated so easily. Define the "hash"-function f from the set of numbers with at most 200 digits to the set of numbers with at most 100 digits by letting $f(x) = (\alpha^{x_0} \beta^{x_1}) \bmod p$, where $x = x_1 q + x_0$. It can now be proven that finding two numbers x and z in the correct interval such that $h(x) = h(z)$ is the same as finding the number a in the equation $(\alpha^a) = \beta$ in $\mathbb{Z}_p$, which is assumed to be hard. One could have defined this function for all natural numbers, but then we would have discovered that the number $x_1(p-1)q + x_0$ and x_0 would be mapped to the same number by the function f, so then the function f would not satisfy the third property we want a hash function to satisfy. This hash function based on the discrete logarithm problem was first proposed by Chaum, van Heijst, and Pfitzmann in 1992.

The algorithms, which are used to calculate hash functions, are usually publicly known. Let us just give the hash function the name hh. hash functions are, for example, used to verify that a received message is valid. Usually, a received encrypted message m is coming together with the message $hh(m)$. The receiver can then calculate $hh(m)$ from m and compare that with the received value $hh(m)$ and see that these two values coincide. This can also be used as an error detecting code in order to ensure that nothing has happened in the transmission.

How can hash functions be used in cryptography? Let us assume that A and B want to send a message to each other over an open line where everybody can read the message. How can A and B communicate without anybody being able to read the message? Let us assume that A and B have a secret key $H(A, B)$ of approximately 100 digits, which only they know. Then A can calculate with help of the hash function a seemingly arbitrarily large "one time pad" in the following way: A applies the hash function on $H(A, B)$, and concatenates the 100 digits he gets to the number $H(A, B)$ and gets a number of 200 digits. Then he evaluates the hash functions on this new number, and gets another 100 digits. He then concatenates these to the last 100-digit numbers to again obtain a 200-digit number on which he can use the hash function. Continuing in this way A gets an arbitrary long sequence of numbers that he can use as a one-time pad. B does the same thing and she also gets the same number as a one-time pad, making it possible for her to decrypt the message from A.

The number $H(A, B)$ which is the secret A and B is sharing, can be exchanged through the RSA or ElGamal key exchange system. This secret number $H(A, B)$ is then used only once and is called a session key. There is also an efficient way of transmitting messages since B can start the decryption before she has gotten the whole message from A.

We will now go back to public key cryptography.

3.3 Public Key Cryptography

Let us now go back to our original starting point with numbers before we started with the digital form of them. We can think of numbers with 100 digits in our usual base 10 number system, and these numbers will from now on be called messages.

Each of the persons A, B, and E chooses a secret that we will call H_A, H_B, and H_E, respectively. (We will come back to how this is done in the RSA and ElGamal system, respectively.) From these secrets, each of the persons A, B, and E can form a couple of functions (a_H, a_P) belonging to A, (b_H, b_P) belonging to B, and (e_H, e_P) belonging to E, respectively. These are also called keys. The functions, or keys, with index P will be published and therefore available for everyone. The functions a_H, b_H, and e_H are called the private functions, or the private keys, and they will be kept strictly secret by the respective owner. The property these functions a_H, b_H, and e_H have to satisfy is that $a_H(a_P(m)) = m$, $b_H(b_P(m)) = m$, and $e_H(e_P(m)) = m$ for all messages m. The knowledge of the functions a_P, b_P, and e_P shall not give any information about how the functions a_H, b_H, and e_H can be calculated, respectively.

If now B wants to send the message m to A, then B takes her message m, applies the public function a_P belonging to A to the message m, and sends the encrypted message $a_P(m)$ to A. A can then use his private function or private key a_H on the received message $a_P(m)$ and get the message m back since $a_H(a_P(m)) = m$. However, neither E nor anybody else knows the function a_H, which is required in order to restore m from $a_P(m)$.

We will soon also include a signature, but let us first give a proper example of how one can organize such a scheme for public key cryptography based on the believed fact that it is hard to find the primes p and q when one knows $n = pq$ and that p and q are prime numbers having more than 150 digits each.

3.4 RSA-Public Key Cryptography

The RSA-system is based on two facts: One from experience, that it is hard to find the prime factorization of a number when the prime factors are large. The second fact is the one we have already mentioned and which goes back to the Swiss mathematician Euler (1707–1783), and that is that $x^{z\phi(n)+1} = x$ in $\mathbb{Z}_n$ for all x in $\mathbb{Z}_n$ and all integers z when n is a square-free positive integer.

We also need an algorithm discovered by Euclid (ca. 300 B.C.), which is also fast in practice, and that is to find, for given positive integers n and m, the unique q and r such that $m = qn + r$ with $0 \leq r \leq n - 1$. This can in turn be used to find the greatest common divisor d of two numbers m and n and express $d = sm + tn$ with integers s and t.

Let us suppose that A picks two prime numbers p_A and q_A of size approximately 10^{150}, which will be the private H_A of A. He multiplies these two primes and obtains the number $n_A = p_A \cdot q_A$, which is a number with about 300 digits. By choosing the primes big enough, A can ensure that each message that has 300 digits is in the interval from 0 to $n_A - 1$. A knows the numbers p_A, q_A and can therefore calculate the number $\phi(n_A) = (p_A - 1)(q_A - 1)$. Those who know the number n_A cannot in a reasonable time find the number $\phi(n_A)$. To find the number $\phi(n_A)$ is equivalent to finding the prime factorization of n_A.

The number $\phi(n_A) = (p_A - 1)(q_A - 1)$ is known to A, but nobody else knows this number. A can therefore also find two numbers t_A and s_A, where s_A can be found by the Euclidean algorithm as soon as t_A is fixed relative prime to $\phi(n_A)$ such that $s_A \cdot t_A = z \cdot \phi(n_A) + 1$ with z, a positive number. A will then make n_A and t_A known to the public, but he keeps the primes p_A and q_A together with the number s_A secret. The function a_P is now defined by $a_P(x) = (x^{t_A}) \bmod n_A$, and the private function or the private key a_H will be given by $a_H(x) = (x^{s_A}) \bmod n_A$, which we will now show will work. We get for all x in the interval 0 to $n_A - 1$ that

$$a_H(a_P(x)) = (x^{t_A} \bmod n_A)^{s_A} \bmod n_A = (x^{t_A \cdot s_A}) \bmod n_A$$

$$= (x^{z \cdot \phi(n_A)+1}) \bmod n_A = x,$$

where one is using the result of Euler in the last equality to obtain this result. For those who do not know the number s_A and therefore have no knowledge of the function a_H, the encrypted message $a_P(m)$ make no sense, and they cannot restore the original message m. (Usually, one is not only transferring the encrypted message $a_p(m)$, but also a hashed version $hh(m)$ of the original message. A can then compare the hash-value on his decrypted message with the transfered hash-value as a control.)

3.5   RSA-Public-Key-Cryptography with Signature

The RSA-scheme for public key cryptography, which we presented above, can also be extended by including a signature ensuring the recipient A that it was actually B who sent the message.

When A has chosen his private H_A, which consists of the prime numbers p_A and q_A, he has calculated n_A and $\phi(n_A)$, chosen the number t_A, and calculated s_A and made n_A and t_A known to the public, and thereby also made the function $a_P(x) = (x^{t_A}) \bmod n_A$ known to the public. Participant B does the same thing, chooses her private B_H consisting of two prime number

p_B and q_B, calculates $n_B = p_B \cdot q_B$ and $\phi(n_B)$, chooses t_B such that t_B and $\phi(n_B)$ are relatively prime, and calculates s_B such that $t_B \cdot s_B = z \cdot \phi(n_B) + 1$ for a positive integer z, and makes n_B and t_B known to the public, and therefore also the function $b_P(x) = (x^{t_B}) \bmod n_B$ known to the public. B would like to send the message m to A in such a way that A and only A can read the message, and in addition A will be able to ensure that the message is coming from B. One can say that B has put her signature on the message.

This can be achieved in the following way: The sender B takes her message m. Let us assume that $n_A \geq n_B$. B then first calculates $b_H(m) = (m^{s_B}) \bmod n_B$ and then $a_P(b_H(m)) = (((m^{s_B}) \bmod n_B)^{p_A}) \bmod n_A$ which B sends to A. A also knows that $n_A \geq n_B$, so he calculates first $a_H(a_P(b_H(m))) = b_H(m)$ by using his private function a_H. Then he uses the function b_P, which is known by everyone from the information n_B and t_B given by B and obtains the result $b_P(b_H(m)) = m$.

With this calculation, the message is correct only if it was encrypted with the private key of B before it was encrypted with the public key of A, and since B is the only one who knows her personal private key, the message is coming from someone who knows the private key of B and that one has to be B. If the inequality goes the other direction, the message has to be broken down into two messages before doing the calculation.

3.6 Problem with Signatures

Will A be convinced that the original message is coming from B and not from E? Let us assume that B is a bit careless and instead of the order of encryption described above, she first encrypts her message with the public key of A, and then encrypts once more with her own private key. E can then read the message $b_H(a_P(m))$, and E can also calculate $b_P(b_H(a_P(m))) = a_P(m)$ and then send $e_H(a_P(m))$ to A. A is then able to reconstruct the message using his own private key in combination with the public key of E, and A is fooled to believe that the message is coming from E.

So, the moral is, sign before using the public key of the recipient. Otherwise, you end up signing "the envelope" and not the document within the envelope. If that is done, everybody who gets the envelope in their hands can just change the envelope and write their own signature on the new envelope without altering the document.

3.7 Receipt

Maybe the sender B in the last section will like to have a receipt from A, ensuring that A has gotten the message. This can be done the following way: B sends the message m to A by sending the signed and encrypted message $a_P(b_H(m))$ to A. A decrypts the message from B and verifies that it is coming from B. Now A can, for example, take the mirror image of the message, encrypt that with his private key, and then encrypt with the public key of B and send the result to B. B decrypts the message and sees that the message she gets is the mirror image of the original message. Only A could have done this, so B is convinced that the message has reached A. This is of course just an example of how this can be done.

3.8 Secret Sharing Based on Discrete Logarithm Problems

A and B want to share a secret session key. Pick a big prime p and a primitive element α, that is, an element α in $\mathbb{Z}_p$ such that $\alpha^c \neq 1$ for all $1 \leq c \leq p - 2$. Now A chooses a secret number a relative prime to $p - 1$ and a random secret number k, and calculates α^k in $\mathbb{Z}_p$, which will be the secret A wants to share with B. A let p and α be known to the public and sends off $\beta = \alpha^{ka}$ to B. B chooses a secret number b relative prime to $p - 1$, calculates $\gamma = \beta^b$, and returns this to A. A returns $\delta = \gamma^c$ to B, where $1 = ca + d(p - 1)$. B then finds e and f such that $1 = eb + f(p - 1)$ and calculates $\delta^e = \alpha^{kabce} = \alpha^{k(1-c(p-1))(1-f(p-1))} = (\alpha^k)^{g(p-1)+1}$ with $g = fc - f - c$. The number δ^e is then by Fermat's theorem equal to α^k. Now both A and B knows α^k, but nobody else is able to find this number unless they are able to find two of the three numbers ak, abk, and bk, which is the same as solving the discrete logarithm problem. The common secret α^k can then be used as a session key for secure communication between A and B.

3.9 Further Reading

We recommend the text book by Douglas R. Stinson [1], where a mathematical background is provided when needed.

Bibliography

[1] D. R. Stinson. *Cryptography: Theory and Practice.* CRC Press, third edition, 2006.

4

Cryptographic Hash Functions

D. Gligoroski

Department of Telematics, NTNU

CONTENTS

This chapter summarizes the development in the field of cryptographic hash functions in the last 20 years. It gives both theoretical foundations as well as a broad list of applications of cryptographic hash functions in numerous algorithms, protocols, and schemes in the field of information security.

4.1   Introduction

The concept of cryptographic hash function is the most useful mathematical concept that has been invented in the contemporary cryptology in the past 20 years. Besides their main use as a part of the digital signatures schemes, there are dozens of other techniques that use the properties of the cryptographic hash functions, and many new algorithms, protocols, and schemes are still being invented.

The basic motivation for constructing hash functions is to build up something that will produce a unique fingerprint for digital files. Loosely speaking, the request for the uniqueness of the fingerprints is actually a twofold request: the cryptographic hash function should be one-way, and it should be collision free. Besides that, the fingerprints should be small enough in order to efficiently store them and to easily manipulate with them. Their size (the hash size, or sometimes called the digest size) can be from 128 up to 512 bits.

The practical requirements for a cryptographic hash function $H()$ can be described by these two (three) requirements:

one-way: The cryptographic hash function $H()$ has to be "one-way" from two perspectives:

> **preimage resistant:** It should have the property that it is "easy" to compute $H(M) = h$ for a given M, but it is "hard" (or "infeasible") to compute M if just the value of h is given.

> **second preimage resistant:** It should have the property that for a given M_1 it is "easy" to compute $H(M_1) = h$, but it is "hard" (or "infeasible") to find another $M_2 \neq M_1$ such that $H(M_2) = H(M_1) = h$.

collision resistance: The cryptographic hash function should be "collision resistant"; that is it should be "hard" (or "infeasible") to find two values $M_1 \neq M_2$ such that $H(M_1) = H(M_2)$.

However, what is the theoretical knowledge that we have that is addressing the above practical requirements?

TABLE 4.1
Theoretical facts or knowledge versus practical requirements for cryptographic hash functions

Theoretical facts	Practical requirements
No one knows whether one-way functions exist, because if they exist, then we will know that P $\neq$ NP. Moreover, since we do not know whether P $\neq$ NP implies that one-way functions exist, proving the existence of one-way functions is even harder than proving that P $\neq$ NP.	Construct a one-way function!
For any function $h : \{0,1\}^* \to \{0,1\}^n$, there exists an infinite number of colliding pairs $M_1, M_2 \in \{0,1\}^*$ such that $h(M_1) = h(M_2)$.	Although there is an infinite number of colliding pairs, practical finding of at least one such a pair should be infeasible!

The first explicit note for the need of one-way functions in cryptography was given by Diffie and Hellman in 1976 [1] and was followed by several significant theoretical results by Yao in 1982 [2] and Levin in 1985 [3], where the existence of one-way functions was connected to the famous question from the complexity theory: "Is P = NP ?"

Namely, it was shown that, if one-way functions exist, then P $\neq$ NP (or vice versa, if P = NP, then there are no one-way functions). And there lies the first controversy: On one hand it is a well-known fact that so far no one has succeeded in neither proving nor disproving the claim P = NP, but on the other, the practical requirements from the designers of cryptographic hash functions are to construct such functions.

The second requirement (for collision resistance) seems also controversial since it is a well-known and trivial theoretical fact that if the domain of the hash function is the set of all possible bit sequences $\{0,1\}^*$ and the range of the hash function is the finite set of n-bit sequences $\{0,1\}^n$, that is, if $H : \{0,1\}^* \rightarrow \{0,1\}^n$, then there is an infinite number of colliding pairs $M_1, M_2 \in \{0,1\}^*$ such that $H(M_1) = H(M_2)$. The controversy of the practical requirements lies in the fact that although there is an infinite number of colliding pairs, the construction of the cryptographic hash function should be such that it is *infeasible in practice* to find at least one such a pair. Table 4.1 summarizes theoretical knowledge that we have versus the practical requirements for the design of cryptographic hash functions. And actually the conflict (or discrepancy) between theoretical knowledge and practical needs is what makes the construction of cryptographic hash functions hard.

The first cryptographer that took the hard task of designing a "cryptographic hash function" was Ron Rivest back in the late 1980s. RSA Data Security, Inc., gave Rivest a task to design a hash function that will be preimage, second preimage, and collision resistant, and he came up with the design of MD2 [4]. Then, in 1990, he designed MD4 [5] and in 1992 MD5 [6]. His designs inspired a whole family of designs now known as the MDx family. This family includes the hash functions HAVAL [7], RIPEMD [8], as well as many others. The historical fact is that as those hash functions were designed, cryptographers were analyzing them and were breaking them.

Then NSA came on the scene and designed the Secure Hash Algorithm (SHA) [9] based on the MDx principles. That function was proposed for standardization via NIST in 1993. While the digest size of MDx hash functions was 128 bits, the size of the SHA was increased to 160 bits. However, after a few years, NSA discovered a weakness in SHA and promptly proposed a tweak. The original SHA is now known as SHA-0 and the tweaked algorithm, as it is known today, is SHA-1 [10].

Aware of the constant progress of the public community in cryptanalysis and breaking the proposed cryptographic hash functions, NSA built a new hash function under the name SHA-2, which NIST adopted as a standard in 2000 [10]. In SHA-2, several new design principles were introduced, and the digest size has increased to 224, 256, 384, or 512 bits.

4.2   Definition of Cryptographic Hash Function

Although we have given an informal definition of cryptographic hash functions in the Introduction, in this section, we will give a more precise and formal definition. There are many equivalent formal mathematical definitions for the three requirements: preimage resistant, second-preimage resistant, and collision resistant. Here we will use a similar style as it was used by Stinson in [11].

Let $\mathcal{F}$ be a family of functions $f \in \mathcal{F}$ such that $f : X \to Y$ where the size of the domain X is $|X| = N$ and the size of the codomain Y is $|Y| = M$. Usually, we assume that $N > M$. Moreover, in the literature, if X is a finite domain and if $f : X \to Y$ is used to define a function over bigger domains than X (in some iterative manner), then f is called *a compression function*. Normally, $\mathcal{F} \subseteq \mathcal{F}^{X,Y}$, where $\mathcal{F}^{X,Y}$ is the set of all functions from X to Y.

Let us use the following assumptions:

1. $f \in \mathcal{F}^{X,Y}$ is chosen uniformly at random.

2. We are given only an oracle access to the function f.

The formal definition of the preimage, second-preimage, and the collision problem for the functions $f \in \mathcal{F}^{X,Y}$ is the following:

Preimage problem: We are given an instance of the function $f : X \to Y$ and an element $y \in Y$. Find an $x \in X$ such that $f(x) = y$.

Second-preimage problem: We are given an instance of the function $f : X \to Y$ and an element $x \in X$. Find an $x' \in X$ such that $x' \neq x$ and $f(x') = f(x)$.

Collision problem: We are given an instance of the function $f : X \to Y$. Find $x, x' \in X$ such that $x' \neq x$ and $f(x') = f(x)$.

In what follows, we will give three algorithms and their computational complexity for solving the mentioned problems. The algorithms $\mathcal{A}_1(q)$, $\mathcal{A}_2(q)$, and $\mathcal{A}_3(q)$ are so-called *randomized algorithms*, where q is the number of queries, that is, invocations of the function f (the oracle access to f). We will denote by ϵ the probability that the randomized algorithm will solve the problem with q queries.

Algorithm $\mathcal{A}_1(f, y, q)$, FINDPREIMAGE

Input: $f \in \mathcal{F}^{X,Y}$, $y \in Y$, q

Choose $X_0 \subseteq X$, $|X_0| = q$
for each $x \in X_0$ do
    if $f(x) = y$ then return(x)
endfor
return(failure)

Algorithm $\mathcal{A}_2(f, x, q)$, FINDSECONDPREIMAGE

Input: $f \in \mathcal{F}^{X,Y}$, $x \in X$, q

Compute $y = f(x)$,
Choose $X_0 \subseteq X \setminus \{x\}$, $|X_0| = q - 1$
for each $x_0 \in X_0$ do
    if $f(x_0) = y$ then return(x_0)
endfor
return(failure)

Algorithm $\mathcal{A}_3(f, q)$, FINDCOLLISION

Input: $f \in \mathcal{F}^{X,Y}$, q

Choose $X_0 \subseteq X$, $|X_0| = q$
for each $x \in X_0$ do
    Compute $y_x = f(x)$
endfor
if $y_x = y_{x'}$ for some $x \neq x'$
    then return(x, x')
    else return(failure)

Without proofs, we will mention the following three theorems:

Theorem 1 ([11]). *For any $X_0 \subseteq X$ with $|X_0| = q$, the success probability of Algorithm* FINDPREIMAGE $\mathcal{A}_1(f, y, q)$ *is*

$$\epsilon = 1 - \left(1 - \frac{1}{M}\right)^q. \tag{4.1}$$

Theorem 2 ([11]). *For any $X_0 \subseteq X \setminus \{x\}$ with $|X_0| = q - 1$, the success probability of Algorithm* FINDSECONDPREIMAGE $\mathcal{A}_2(f, y, q)$ *is*

$$\epsilon = 1 - \left(1 - \frac{1}{M}\right)^{q-1}. \tag{4.2}$$

Theorem 3 ([11]). *For any $X_0 \subseteq X$ with $|X_0| = q$, the success probability of Algorithm* FINDCOLLISION $\mathcal{A}_3(f, q)$ *is*

$$\epsilon = 1 - \left(\frac{M-1}{M}\right)\left(\frac{M-2}{M}\right)\cdots\left(\frac{M-q+1}{M}\right). \tag{4.3}$$

We want to note that if we set the success probability ϵ for the FINDCOLLISION algorithm to be at least 0.5, that is, $\epsilon \geq 0.5$, and if we can express the number of elements M in the set Y as $M = 2^n$ for some integer value n, then solving the expression (4.3) for the number of queries q will give us the following expression:

$$q \approx 1.18 \times 2^{\frac{n}{2}}. \tag{4.4}$$

The expression (4.4) is the well-known birthday paradox bound for finding a collision of a random function $f : X \to Y$ with the codomain Y that has 2^n elements.

Finally, we have enough theoretical material to define a function that we can call a *cryptographic hash function*.

Definition 1. We call the function $f : X \to Y$ a **cryptographic hash function**, where the size of the domain X is $|X| = N$, the size of the codomain Y is $|Y| = M$, and $N > M$, if we do not know algorithms for solving the problems FINDPREIMAGE, FINDSECONDPREIMAGE, and FINDCOLLISION that are faster than algorithms $\mathcal{A}_1(f, y, q)$, $\mathcal{A}_2(f, y, q)$, and $\mathcal{A}_3(f, q)$, that is, that have bigger success probabilities than the probabilities described in expressions (4.1), (4.2), and (4.3).

Note that the Definition 1 is pretty subjective, since *it refers to our lack of knowledge for solving certain problems.* But having in mind the history of design, development, and cryptanalysis of cryptographic hash functions, we think that actually this definition captures the essence of the functions that once we classified as cryptographic hash functions (MD4, MD5, SHA-0, SHA-1), but recent cryptanalytic developments have put them out of that class (i.e., in the class of non-cryptographic or broken cryptographic hash functions).

For a more extensive elaboration about the controversies and difficulties connected to the definition of collision-resistant hash functions and the ways to formalize our lack of knowledge of algorithms that will solve the FINDCOLLISION problem faster than the algorithm $\mathcal{A}_3(f, q)$, we recommend the paper of Rogaway from 2007: "Formalizing Human Ignorance: Collision-Resistant Hashing without the Keys" [12].

4.3 Iterated Hash Functions

4.3.1 Strengthened Merkle-Damgård Iterated Design

Almost all practical designs of hash functions are based on iteration of a compression function (as generally defined in Section 4.2) with arbitrary but fixed size input that is usually much bigger than the output of the compression function. A typical technique for implementing such an iterative design is by partitioning the input message into blocks of fixed length called block length.

However, since the length of the message can be arbitrary, in order for the message to be processed properly by the hash function it needs to be properly padded such that the length of the padded message is a multiple of the block length. This padding ought to be an injective mapping to a message with a bigger length, since we want to avoid producing trivial collisions with non-injective padding procedures. Although there are many injective schemes for the message padding, one that is most known is the one that appends the bit "1" to the message, then appends (possibly many) bits "0", and at the end, it appends the binary presentation of the original message length. This specific injective padding is called Merkle-Damgård strengthening because in 1989, approximately at the same time Merkle [13] and Damgård [14] came to the conclusion that if the compression function f is collision-resistant and if padded with that injective manner, then the whole iterated hash function construction will be collision resistant.

The generic procedure for the strengthened iterated Merkle-Damgård hash design is described in Table 4.2, and its graphic presentation is given in Figure 4.1.

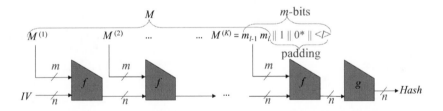

FIGURE 4.1
A graphical presentation of the strengthened Merkle-Damgård iterated hash design.

4.3.2 Hash Functions Based on Block Ciphers

Since the techniques for designing block ciphers were developed by cryptographers a long time before the problem of designing a hash function appeared, a

TABLE 4.2

A generic description of the strengthened Merkle-Damgård iterated hash design

Generic Merkle-Damgård construction of iterated hash function $H()$ with a compression function $f : \{0,1\}^{n+m} \to \{0,1\}^n$
Input: Message M of length l bits, and message digest size n. **Output:** A message digest $Hash$ that is n bits long.

1. Preprocessing

   (a) From M produce a padded message M' such that the length of the padded message M' is multiple of m, that is, $|M'| = K \times m$ for some $K > 0$. Perform the padding such that $M' = M||1||0^*||\langle l \rangle$, where $||$ denotes a concatenation, 0^* denotes an empty or nonempty string of 0s and $\langle l \rangle$ is a binary presentation of the value l.

   (b) Parse the padded message into K m-bit message blocks, $M^{(1)}, M^{(2)}, \ldots, M^{(K)}$.

   (c) Set the initial hash value $IV = H^{(0)}$.

2. Iterated message digestion

   For $i = 1$ to K

   {

   $\qquad H^{(i)} = f(M^{(i)}, H^{(i-1)});$

   }

3. Finalization

   $Hash = g(H^{(K)})$ (the function $g()$ can be some arbitrary function that is usually the same as $f()$ but with different inputs).

natural development was to investigate how a collision-resistant compression function can be built using a block cipher $E_k(x)$, where the block cipher $E()$ maps n-bit messages x to n-bit ciphertexts and is parameterized by n-bit key k, that is, it is a mapping $E : \{0,1\}^n \times \{0,1\}^n \to \{0,1\}^n$.

The general scheme can be described by the following expression:

$$f(h, m) = E_k(x) \oplus s, \text{ where } k, x, s \in \{c, h, m, m \oplus h\}, \qquad (4.5)$$

and where c is a fixed n-bit constant.

In 1993, Preneel, Govaerts, and Vandewalle in [15] located 12 combinations of the parameters $k, x, s \in \{c, h, m, m \oplus h\}$ that can give a collision-resistant compression function. Ten years later, Black et al., [16] proved that those schemes are collision-resistant in the ideal cipher model. Those 12 secure schemes are given in Table 4.3. In the literature, three of those schemes (Nr. 1, 2, and 5) are known by the authors that have first proposed them, and their names are given in the third column of the table.

TABLE 4.3
The 12 PGV schemes that can construct a collision-resistant compression function from a block cipher

Nr.	$f(h, m)$	Note
1.	$E_h(m) \oplus m$	Matyas-Meyer-Oseas scheme
2.	$E_h(m) \oplus m \oplus h$	Miyaguchi-Preneel scheme
3.	$E_h(m \oplus h) \oplus m$	
4.	$E_h(m \oplus h) \oplus m \oplus h$	
5.	$E_m(h) \oplus h$	Davies-Meyer scheme
6.	$E_m(h) \oplus m \oplus h$	
7.	$E_m(m \oplus h) \oplus h$	
8.	$E_m(m \oplus h) \oplus m \oplus h$	
9.	$E_{m \oplus h}(h) \oplus h$	
10.	$E_{m \oplus h}(h) \oplus m$	
11.	$E_{m \oplus h}(m) \oplus h$	
12.	$E_{m \oplus h}(m) \oplus m$	

4.3.3   Generic Weaknesses of the Merkle-Damgård Design

The Merkle-Damgård iterated construction of hash functions suffers from several generic weaknesses not present in an ideal random function that maps strings of arbitrary length to strings of n-bits.

Message Expansion Attack

The first weakness is its vulnerability to the "Length-extension attack". The generic length extension attack on an iterated hash function based on the Merkle-Damgård iterative design principles works as follows.

Let $M = M_1 || M_2 || \dots || M_K$ be a message consisting of exactly K blocks that will be iteratively digested by some compression function $f(A, B)$ according to the Merkle-Damgård iterative design principles, where A and B are messages (input parameters for the compression function) that have the same length as the final message digest. Let P_M be the last (the padding block) of M obtained according to the Merkle-Damgård strengthening. Then, the digest H of the message M, is computed as

$$H(M) = f(\dots f(f(IV, M_1), M_2) \dots, P_M),$$

where IV is the initial fixed value for the hash function.

Now suppose that the attacker does not know the message M, but knows (or can easily guess) the length of the message M. The attacker knows the padding block P_M. Now, the attacker can construct a new message $M' = P_M \| M'_1$ such that he knows the hash digest of the message $M \| M'$, that is,

$$H(M \| M') = f(f(H(M), M'_1), P_{M'}),$$

where $P_{M'}$ is the padding (Merkle-Damgård strengthening) of the message $M \| M'$.

Thus, the attacker knows the hash value of arbitrary many messages $M \| M'$ that have the message M as the first part, without knowing the message M. This can be a potential risk in identification protocols that use a secret prepending M.

Multi-collision Attack

In 2004, Joux [17] found another property of the Merkle-Damgård design that is not present as a property of an ideal random function (random oracle).

Namely, if we are interested in finding 2^r collisions for an ideal random function, the total amount of computational work is $\Theta(2^{\frac{nr(r-1)}{2}})$ calls to the compression function. On the other hand, for the Merkle-Damgård design finding 2^r collisions is possible "with just" $r2^r$ calls to the compression function with the following algorithm:

TABLE 4.4
The multicollision attack of Joux on the Merkle-Damgård iterated hash design

A multicollision attack on the Merkle-Damgård construction of iterated hash function $H()$ with a compression function $f : \{0,1\}^{n+m} \to \{0,1\}^n$
Input: Value r.   **Output:** 2^r colliding messages.
1. Set $IV = h_0$ 2. For $i = 0$ to $r - 1$ By computing $\approx 2^{\frac{n}{2}}$ compression functions $f(h_i, m_i)$, find a colliding pair $(m_i^{(0)}, m_i^{(1)})$ such that $$h_{i+1} = f(h_i, m_i^{(0)}) = f(h_i, m_i^{(1)})$$ 3. Any of the 2^r message compositions $m_0^{(c)} \| m_1^{(c)} \| \ldots \| m_{r-1}^{(c)}$ where $(c) \in \{0, 1\}$ gives a collision.

Herding Attack

In 2006, Kelsey and Kohno in [18] found yet another property of the Merkle-Damgård iterated hash design that allows the attacker to find a message that

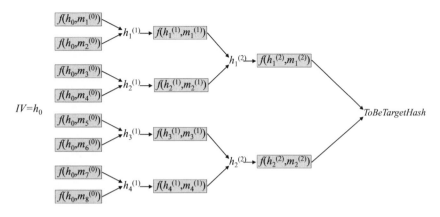

FIGURE 4.2
A graphical presentation of the herding attack.

is a preimage of a predetermined hash value with much less than 2^n calls to the compression function.

Their attack is a kind of a time-memory trade-off attack, and its schematic presentation is given in Figure 4.2. The attacker is building a binary tree with a widest leaf basis of 2^k colliding blocks $m_i^{(0)}, i = 1, \ldots, 2^k$ all starting with the initial value $h_0 = IV$ giving the collisions $h_i^{(1)}, i = 1, \ldots, 2^{k-1}$. Then, subsequently starting from those intermediate values $h_i^{(1)}$ the next level of colliding blocks $m_i^{(1)}, i = 1, \ldots, 2^{k-2}$ is found. This gives the collisions $h_i^{(2)}, i = 1, \ldots, 2^{k-3}$. The procedure continues up to the top level where the colliding value is denoted as "*ToBeTargetHash*".

Note that due to the birthday paradox, if the compression function $f()$ is an ideal random function, finding a collision will take approximately $2^{\frac{n}{2}}$ computations of the compression function, and thus the total computational effort for building the whole tree of $2^k - 1$ colliding values will take approximately $2^{\frac{n}{2}+k}$ calls to the compression function. The authors of [18] gave an algorithm that reduces this number of calls to the compression function even more.

Then the attacker can publish the value of *ToBeTargetHash* claiming that he knows the outcome of some future event (for example, the state of the stock exchange market at December 31, 2020), and that he had hashed the values of the most important stocks at that day and that the hash value of that information P is *ToBeTargetHash*.

When the actual date will come, he will compose the concrete information P, and then he will try to find a message block L such that a proper iterative hashing (but without the proper padding) $H(P||L) = h_i^{(j)}$ will give a value from the tree structure. The computational effort for finding the value L having in mind that there are $2^k - 1$ precomputed hash values $h_i^{(j)}$ is 2^{n-k}. Once the attacker finds that "linking" value $h_i^{(j)}$ he will continue to add message blocks according to the values of the tree up to the top of the tree and the value *ToBeTargetHash*. If the hash function has a padding scheme that adds the length of the original message at the end, this attack will not work directly, but the authors of [18] show how to overcome the padding part so the attack still works.

In the described way the attacker can find a message $P||L||S$ that gives a predetermined hash value, that is, $H(P||L||S) = ToBeTargetHash$ with much less calls to the compression function than the expected 2^n calls.

4.3.4 Wide Pipe (Double Pipe) Constructions

In 2005, Lucks [19, 20] and Coron et al. [21] proposed several changes in the Merkle-Damgård design in order to address the weaknesses in that design. One of the proposed changes was to increase the internal state (the chaining pipe) to be at least two times wider than the final hash digest size. This design avoids the weaknesses against the generic attacks of Joux [17] and Kelsey and Schneier [22], thereby guaranteeing resistance against a generic multicollision attack and length extension attacks. The graphical presentation of the double-pipe design is shown in Figure 4.3.

The design concept of the double-pipe hash construction has attracted big interest from the cryptographic hash designers and as a result the majority of the proposed designs in the SHA-3 competition were double-pipe designs.

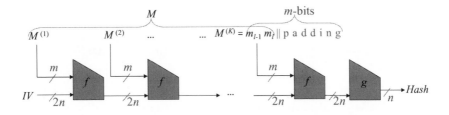

FIGURE 4.3
A graphical presentation of the Double-pipe iterated hash design.

4.3.5 HAIFA Construction

In 2006, guided by the motivation to improve the Merkle-Damgård iterative design of the hash functions, Biham and Dunkelman proposed A Framework for Iterative Hash Functions: HAIFA [23]. The idea is to keep the narrow-pipe construction as it is in the Merkle-Damgård design, but the compression function to receive as inputs additional parameters like *salt* and the number of digested bits so far during the iterative process (denoted as bh_0, bh_1, ..., in the graphical presentation of the HAIFA design that is shown in Figure 4.4).

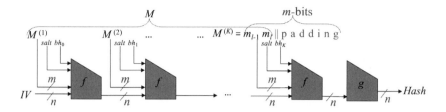

FIGURE 4.4
A graphical presentation of the HAIFA iterated hash design.

4.3.6   Sponge Functions Constructions

Hash function designs based on sponge functions are recent design concepts invented in 2007 by Bertoni, Daemen, Peeters, and Van Assche [24]. This design concept has also attracted big interest from cryptographic hash designers and in the Second Round of the SHA-3 competition there were four sponge (or sponge-like) designs.

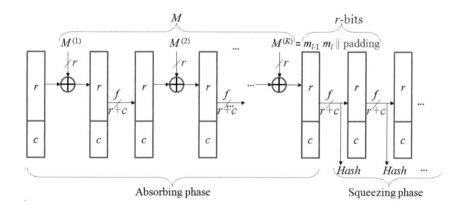

FIGURE 4.5
A graphical presentation of the sponge iterated hash design.

The idea is to use a "random" keyless permutation (or function) f that will map a wider internal state of $r + c$ bits to itself, and to absorb the message in chunks of r bits that are injected by xoring with the first r bits of the internal state. That phase is called *Absorbing phase*. Then, in the *Squeezing phase* the function f iteratively maps the internal state without any further external input, and extracts certain number of bits from the internal state until the required number of bits is achieved.

The graphical presentation of the sponge design is shown in Figure 4.5.

The previous three designs (wide pipe, HAIFA and sponge) do not suffer from the generic weaknesses the Merkle-Damgård design has.

4.4 Most Popular Cryptographic Hash Functions

4.4.1 MD5

MD5 is one in a series of message digest algorithms designed by Rivest [6]. When analytic work indicated that MD5's predecessor MD4 was likely to be insecure, MD5 was designed in 1991 to be a secure replacement. (Weaknesses were indeed later found in MD4 by Hans Dobbertin.)

The compression function of MD5 uses four Boolean functions

$$F(U, V, W) = (U \wedge V) \vee (\neg U \wedge W),$$
$$G(U, V, W) = (U \wedge W) \vee (V \wedge \neg W),$$
$$H(U, V, W) = U \oplus V \oplus W,$$
$$I(U, V, W) = (U \vee \neg W) \oplus V.$$

It uses 64 additive constants, and it updates the value of the variable a (which is one of four working 32-bit variables $a, b, c,$ and d) by one of the four assignments $FF(a, b, c, d, Z, Y, s)$, $GG(a, b, c, d, Z, Y, s)$, $HH(a, b, c, d, Z, Y, s)$, and $II(a, b, c, d, Z, Y, s)$ that are defined as follows:

$$
\begin{aligned}
FF &: a := b + (a + F(b, c, d) + Z + Y)^{\lll s}, \\
GG &: a := b + (a + G(b, c, d) + Z + Y)^{\lll s}, \\
HH &: a := b + (a + H(b, c, d) + Z + Y)^{\lll s}, \\
II &: a := b + (a + I(b, c, d) + Z + Y)^{\lll s}.
\end{aligned}
$$

MD5 has 4 rounds of 16 steps each, and in each step one of the working 32-bit variables is updated.

The successful series of attacks on the MD5 hash function started early in 1993 with the work of den Boer and Bosselaers in their paper [25]. Again, with clever techniques of choosing "magic" numbers as a target values to obtain, they analyze the first two rounds of MD5 finding collisions by "walking forward" and "walking backward", that is, tracing the obtained differences, and solving simple linear equations for some parts of the processed message. The part 2.2 of their algorithm explicitly describes equations that are obtained directly from the equations of MD5. That part involves invertible operations of addition modulo 2^{32} and left rotation.

In 1996 at the rump session of Eurocrypt '96 using similar techniques as den Boer and Bosselaers, Dobbertin provided even better results on finding collisions for MD5. Again, the crucial point was that setting up equations for tracing the differences, although complicated, was possible because the compression function of MD5 uses the invertible operations of addition modulo 2^{32} and left rotation.

Finally, in 2004 at the rump session of CRYPTO 2004 (as well as at the cryptology ePrint archive) Wang et al. presented concrete results of breaking MD5, HAVAL-128, and RIPEMD [26].

In 2006 Klima published an algorithm [27] that can find a collision within one minute on a single notebook computer, using a method he calls tunneling.

4.4.2   SHA-1

We give here a brief description of the main components in the definition of SHA-1. It uses the same three Boolean functions as MD4, but in 4 rounds and in the following order:

$$F(U, V, W) = (U \wedge V) \vee (\neg U \wedge W),$$
$$H(U, V, W) = U \oplus V \oplus W,$$
$$G(U, V, W) = (U \wedge V) \vee (U \wedge W) \vee (V \wedge W),$$
$$H(U, V, W) = U \oplus V \oplus W.$$

It uses four additive constants and has five working 32-bit variables: a, b, c, d, and e. Every round has 20 steps, and it uses an internal procedure for message expansion from 16 to 80 32-bit variables. Its output is 160 bits. The assignments of working variables is done by the following functions:

$$t := a^{\lll 5} + F(b, c, d) + e + W + K,$$
$$(a, b, c, d, e) := (t, a, b^{\lll 30}, c, d),$$

$$t := a^{\lll 5} + H(b, c, d) + e + W + K,$$
$$(a, b, c, d, e) := (t, a, b^{\lll 30}, c, d),$$

$$t := a^{\lll 5} + G(b, c, d) + e + W + K,$$
$$(a, b, c, d, e) := (t, a, b^{\lll 30}, c, d),$$

$$t := a^{\lll 5} + H(b, c, d) + e + W + K,$$
$$(a, b, c, d, e) := (t, a, b^{\lll 30}, c, d),$$

where W are 32-bit variables obtained by message expansion and K can have one of four predefined values of additive constants.

Wang et al. in February 2005 gave a note about their latest findings about SHA-1 and the possibility of finding collisions by 2^{69} SHA-1 hash computations [28]. Although the approach of Wang et al. is much broader and more complicated, we can say that again the basic principle of finding the collisions is exploiting the invertible and linear properties of the functions used in the MD4 family of hash functions.

4.4.3   SHA-2

The SHA-2 family of hash functions was designed by NSA and adopted by NIST in 2000 as a standard that replaced SHA-1 in 2010 [29].

SHA-2 is actually a family of four hash functions with outputs of 224, 256, 384, and 512 bits, and accordingly, sometimes SHA-2 functions are denoted as SHA-224, SHA-256, SHA-384, and SHA-512. The full description of SHA-2 family can be found in [29].

The main difference between those four functions is that SHA-224 and SHA-256 are defined by operations performed on 32-bit variables, while SHA-384 and SHA-512 are defined by operations performed on 64-bit variables.

We give here the definitions of four S-boxes (or bijective transformations) present in the design of SHA-2 that act either on 32 or 64 bits, while the rest of the design specifics can be found in [29].

For SHA-224/256, those four bijective transformations are defined as

$$
\begin{aligned}
\Sigma_0^{256}(x) &= ROTR^2(x) &\oplus& ROTR^{13}(x) &\oplus& ROTR^{22}(x), \\
\Sigma_1^{256}(x) &= ROTR^6(x) &\oplus& ROTR^{11}(x) &\oplus& ROTR^{25}(x), \\
\sigma_0^{256}(x) &= ROTR^7(x) &\oplus& ROTR^{18}(x) &\oplus& SHR^3(x), \\
\sigma_1^{256}(x) &= ROTR^{17}(x) &\oplus& ROTR^{19}(x) &\oplus& SHR^{10}(x),
\end{aligned}
\tag{4.6}
$$

where $ROTR^n(x)$ means rotation of the 32-bit variable x to the right for n positions and $SHR^n(x)$ means shifting of the 32-bit variable x to the right for n positions.

For SHA-384/512, the four bijective transformations are defined as

$$
\begin{aligned}
\Sigma_0^{512}(x) &= ROTR^{28}(x) &\oplus& ROTR^{34}(x) &\oplus& ROTR^{39}(x), \\
\Sigma_1^{512}(x) &= ROTR^{14}(x) &\oplus& ROTR^{18}(x) &\oplus& ROTR^{41}(x), \\
\sigma_0^{512}(x) &= ROTR^1(x) &\oplus& ROTR^8(x) &\oplus& SHR^7(x), \\
\sigma_1^{512}(x) &= ROTR^{19}(x) &\oplus& ROTR^{61}(x) &\oplus& SHR^6(x),
\end{aligned}
\tag{4.7}
$$

where $ROTR^n(x)$ means rotation of the 64-bit variable x to the right for n positions and $SHR^n(x)$ means shifting of the 64-bit variable x to the right for n positions.

Since the design principles behind SHA-2 construction are not publicly available, several public papers produced by the academic cryptographic community have been devoted to analysis of the SHA-2 design and to the cryptanalysis of SHA-2 hash functions. Gilbert and Handschuh in 2003 made an analysis of the SHA-2 family [30]. They proved that there exist XOR-differentials that give a 9-round local collision with probability 2^{-66}. In 2004, Hawkes, Paddon, and Rose [31] improved the result and showed existence of addition-differentials of 9-round local collisions with probability of 2^{-39}. Different variants of SHA-256 have been analyzed in 2005 by Yoshida and Biryukov [32] and by Matusiewicz et al. [33]. In 2006, Mendel et al. [34] found XOR-differentials for 9-round local collisions, also with probability 2^{-39} (recently improved to the value 2^{-38} [35]). In 2008, Nikolić and Biryukov found collisions in 21 step reduced SHA-256, and their attack was afterwards improved by Indesteege et al., up to 24 steps [36].

Previously, an interest to analyze the S-boxes in SHA-2 was described in the work of Matusiewicz et al. [33], where they noted:

- *"The substitution boxes Σ_0 and Σ_1 constitute the essential part of the hash function and fulfil two tasks: they add bit diffusion and destroy the ADD-linearity of the function."*

- *"σ_0 and σ_1 have both the property to increase the Hamming weight of low-weight inputs. This increase is upper bounded by a factor of 3. The average increase of Hamming weight for low-weight inputs is even higher if three*

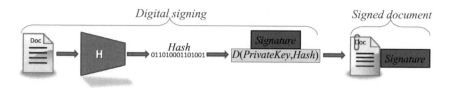

FIGURE 4.6

A graphical presentation of the signing process of the document "Doc" with a private key $PrivateKey$ using the cryptographic hash function H.

rotations are used instead of two rotations and one bit-shift. However, a reason for this bit-shift is given by the next observation."

- *"In contrast to all other members of the MD4-family including SHA-1, rotating expanded message words to get new expanded message words is not possible anymore (even in the XOR-linearised case). This is due to the bit-shift being used in σ_0 and σ_1."*

4.4.4 NIST SHA-3 Hash Competition

Following the development in the field of cryptographic hash functions, NIST organized two cryptographic hash workshops [37] in 2005 and 2006, respectively. As a result of those workshops, NIST decided to run a 4 year worldwide open hash competition for selection of the new cryptographic hash standard SHA-3 [38]. The requirements for the hash digest size for the new cryptographic hash functions are 224, 256, 384, and 512 bits, the same as for the current SHA-2 standard. Out of 64 initial submissions, 51 entered the First Round [39], 14 have been selected for the Second Round of the SHA-3 competition [40], and 5 are now selected for the Final Round [41].

4.5 Application of Cryptographic Hash Functions

4.5.1 Digital Signatures

The most important use of cryptographic hash functions is in the production and verification of digital signatures. The paradigm of public key cryptography proposed by Diffie and Hellman in 1976 in [1] brought two new important concepts in the modern cryptology: (1) A secure communication can be established between two parties even if they did not exchange one secret, shared, and symmetric key. (2) A cryptographic scheme can imitate all the legal aspects of the written signature but applied to a digital content.

TABLE 4.5

Two essential parts of the digital signatures: Signing and Verification process

Signing	Verification
Input: Document Doc, Hash function H, Private Key $PrivateKey$	**Input:** Signed document $Doc\|Signature$, Hash function H, Public Key $PublicKey$
1. $Hash = H(Doc)$ 2. $Signature = D(PrivateKey, Hash)$ 3. **Output:** $Doc\|Signature$	1. $Hash_1 = H(Doc)$ 2. $Hash_2 = E(PublicKey, Signature)$ 3. **Output:** If $Hash_1 == Hash_2$, **TRUE**, else **FALSE**

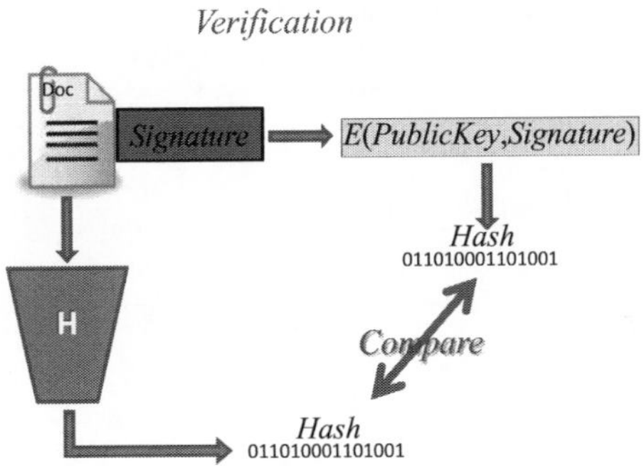

FIGURE 4.7

A graphical presentation of the verification process of the signed document "Doc || Signing" with the public key $PublicKey$ using the cryptographic hash function H.

In the whole concept of digital signatures, the role of the cryptographic hash functions is to produce a message digest from the document that is a subject of signing, and then operations with the private key to be performed on the produced hash value. Then, in the process of verification, the document that is accompanied by its digital signature is hashed, the digital signature is transformed by the public key of the signer, and if those two results are same, the digital signature is verified.

The formal mathematical modeling of the signing and verification process is given in Table 4.5, and the schematic presentation of the process of signing and verification is presented in Figure 4.6 and in Figure 4.7.

TABLE 4.6

A list of applications where hash functions are used

Application in	Used hash functions	Use frequency	Application in	Used hash functions	Use frequency
Digital signatures	MD5 SHA-1 SHA-256 SHA-512	Rare High Rare Rare	Data Integrity	MD5 SHA-1 SHA-256 SHA-512	High Modest Rare Rare
Commitment schemes	MD5 SHA-1 SHA-256 SHA-512	Modest High Rare Rare	Password protection	MD5 SHA-1 SHA-256 SHA-512	High High Modest Rare
Microsoft CLR strong names	MD5 SHA-1 SHA-256 SHA-512	None High Rare Rare	Python setuptools	MD5 SHA-1 SHA-256 SHA-512	High High Modest Rare
Software packet managers	MD5 SHA-1 SHA-256 SHA-512	High High Modest Rare	Google micro-payment system	MD5 SHA-1 SHA-256 SHA-512	None High None None
Security mechanism for					
Local file systems	MD5 SHA-1 SHA-256 SHA-512	High High Rare Rare	Decentralized file systems	MD5 SHA-1 SHA-256 SHA-512	High High Rare Rare
P2P file-sharing	MD5 SHA-1 SHA-256 SHA-512	High High Rare Rare	Decentralized revision control tools	MD5 SHA-1 SHA-256 SHA-512	High High Rare Rare
Intrusion detection systems	MD5 SHA-1 SHA-256 SHA-512	High High Rare Rare	De-duplication systems	MD5 SHA-1 SHA-256 SHA-512	High High Rare Rare

4.5.2 Other Applications

Hash functions are used to provide data integrity verification, in commitment schemes, and in password protection at the server side. They are also used as a basic security mechanism for local file systems, decentralized file systems, P2P file-sharing, in decentralized revision control tool, and have many other applications. In Table 4.6 we give an extensive (but far from complete) list of applications of cryptographic hash functions. The list was composed from diverse Internet sources, cryptographic forums and the NIST hash mailing list hash-forum@nist.gov.

4.6   Further Reading and Web Sites

Handbook of Applied Cryptography [42] is a well known book in this CRC series. Security in Computing [43] is a comprehensive textbook on this topic. The basis of discrete mathematics can be found in Rosen's handbook [44]. The International Association for Cryptologic Research provides a web site [45] for ongoing cryptology research activities.

Bibliography

[1] W. Diffie and M. Hellmann. New directions in cryptography. In *IEEE Trans. on Info. Theory*, volume IT-22, pages 644–654, 1976.

[2] A. Yao. Theory and application of trapdoor functions. In *Proceedings of 23rd IEEE Symposium on Foundations of Computer Science*, pages 80–91, 1982.

[3] L. A. Levin. One-way functions and pseudorandom generators. *Combinatorica*, 7(4):357–363, 1987.

[4] B. S. Kaliski. The MD2 message-digest algorithm. In *RFC 1319*. Network Working Group, RSA Laboratories, April 1992.

[5] R. L. Rivest. The MD4 message-digest algorithm. In *RFC 1186*. Network Working Group, MIT Laboratory for Computer Science and RSA Data Security Inc., October 1990.

[6] R. L. Rivest. The MD5 message-digest algorithm. In *RFC 1321*. Network Working Group, MIT Laboratory for Computer Science and RSA Data Security Inc., April 1992.

[7] Yuliang Zheng, Josef Pieprzyk, and Jennifer Seberry. Haval – A one-way hashing algorithm with variable length of output (extended abstract). In Jennifer Seberry and Yuliang Zheng, editors, *Advances in Cryptology – AUSCRYPT '92*, volume 718 of *Lecture Notes in Computer Science*, pages 81–104. Springer Berlin / Heidelberg, 1993.

[8] A. Bosselaers, H. Dogbbertin, and B. Preneel. The RIPEMD-160 cryptographic hash function. 22(1):24, 26, 28, 78, 80, January 1997.

[9] National Institute of Standards and Technology (NIST). *Publication YY: Announcement and Specifications for a Secure Hash Standard (SHS)*, January 22 1992.

[10] NIST. *Secure Hash Standard.* National Institute of Standards and Technology, Washington, 2002. Federal Information Processing Standard 180-2.

[11] D. R. Stinson. Some observations on the theory of cryptographic hash functions. *Des. Codes Cryptography,* 38:259–277, February 2006.

[12] P. Rogaway. Formalizing human ignorance. In Phong Q. Nguyen, editor, *VIETCRYPT,* volume 4341 of *Lecture Notes in Computer Science,* pages 211–228. Springer, 2006.

[13] Ralph C. Merkle. One way hash functions and DES. In Gilles Brassard, editor, *CRYPTO,* volume 435 of *Lecture Notes in Computer Science,* pages 428–446. Springer, 1989.

[14] Ivan Damgård. A design principle for hash functions. In Gilles Brassard, editor, *CRYPTO,* volume 435 of *Lecture Notes in Computer Science,* pages 416–427. Springer, 1989.

[15] B. Preneel, R. Govaerts, and J. Vandewalle. Hash functions based on block ciphers: A synthetic approach. In *Proceedings of CRYPTO 1993,* volume 773 of *LNCS,* pages 368–378, 1994.

[16] J. Black, P. Rogaway, and T. Shrimpton. Black-box analysis of the block-cipher-based hash function constructions from PGV. In *Proceedings of CRYPTO 2002,* volume 2442 of *LNCS,* pages 320–335, 2002.

[17] A. Joux. Multicollisions in iterated hash functions. application to cascaded constructions. In *Proceeding of CRYPTO 2004,* volume 3152 of *LNCS,* pages 430–440, 2004.

[18] John Kelsey and Tadayoshi Kohno. Herding hash functions and the Nostradamus attack. In Serge Vaudenay, editor, *EUROCRYPT,* volume 4004 of *Lecture Notes in Computer Science,* pages 183–200. Springer, 2006.

[19] S. Lucks. Design principles for iterated hash functions. Cryptology ePrint Archive, Report 2004/253, 2004. http://eprint.iacr.org/.

[20] S. Lucks. A failure-friendly design principle for hash functions. In *Proceeding of ASIACRYPT 2005,* volume 3788 of *LNCS,* pages 474–494, 2005.

[21] J.-S. Coron, Y. Dodis, C. Malinaud, and P. Puniya. Merkle–Damgård revisited: How to construct a hash function. In *Advances in Cryptology: CRYPTO 2005,* volume 3621 of *LNCS,* pages 430–440. Springer-Verlag, 2005.

[22] J. Kelsey and B. Schneier. Second preimages on n-bit hash functions for much less than 2^n work. In *Proceeding of EUROCRYPT 2005,* volume 3494 of *LNCS,* pages 474–490, 2005.

[23] E. Biham and O. Dunkelman. A framework for iterative hash functions – HAIFA. In *Second NIST Cryptographic Hash Workshop*, volume 2006, page 2, 2006.

[24] Guido Bertoni, Joan Daemen, Michael Peeters, and Gilles Van Assche. Sponge functions, 2007. http://csrc.nist.gov/groups/ST/hash/documents/JoanDaemen.pdf.

[25] Bert den Boer and Antoon Bosselaers. Collisions for the compressin function of MD5. In *EUROCRYPT*, pages 293–304, 1993.

[26] Xiaoyun Wang, Dengguo Feng, Xuejia Lai, and Hongbo Yu. Collisions for hash functions MD4, MD5, HAVAL-128 and RIPEMD. Cryptology ePrint Archive, Report 2004/199, 2004. http://eprint.iacr.org/.

[27] Vlastimil Klima. Tunnels in hash functions: MD5 collisions within a minute. Cryptology ePrint Archive, Report 2006/105, 2006. http://eprint.iacr.org/.

[28] X. Wang, Y. L. Yin, and H. Yu. Collision search attacks on SHA1. White paper, 2005. http://www.c4i.org/erehwon/shanote.pdf, accessed 1 April 2011.

[29] NIST. *Secure Hash Standard*. National Institute of Standards and Technology, Washington, 2002. Federal Information Processing Standard 180-2.

[30] Henri Gilbert and Helena Handschuh. Security analysis of SHA-256 and sisters. In Mitsuru Matsui and Robert J. Zuccherato, editors, *Selected Areas in Cryptography*, volume 3006 of *Lecture Notes in Computer Science*, pages 175–193. Springer, 2003.

[31] Philip Hawkes, Michael Paddon, and Gregory G. Rose. On corrective patterns for the SHA-2 family. Cryptology ePrint Archive, Report 2004/207, 2004. http://eprint.iacr.org/.

[32] Hirotaka Yoshida and Alex Biryukov. Analysis of a SHA-256 variant. In Bart Preneel and Stafford E. Tavares, editors, *Selected Areas in Cryptography*, volume 3897 of *Lecture Notes in Computer Science*, pages 245–260. Springer, 2005.

[33] Krystian Matusiewicz, Josef Pieprzyk, Norbert Pramstaller, Christian Rechberger, and Vincent Rijmen. Analysis of simplified variants of SHA-256. In Christopher Wolf, Stefan Lucks, and Po-Wah Yau, editors, *WEWoRC*, volume 74 of *LNI*, pages 123–134. GI, 2005.

[34] Florian Mendel, Norbert Pramstaller, Christian Rechberger, and Vincent Rijmen. Analysis of step-reduced SHA-256. In Matthew J. B. Robshaw, editor, *FSE*, volume 4047 of *Lecture Notes in Computer Science*, pages 126–143. Springer, 2006.

[35] M. Hölbl, C. Rechberger, and T. Welzer. Finding message pairs conforming to simple SHA-256 characteristics. In Christopher Wolf Stefan Lucks, Ahmad-Reza Sadeghi, editor, *Proceeding Records of WEWoRC 2007 - Western European Workshop on Research in Cryptology*, pages 21–25, 2007.

[36] Sebastiaan Indesteege, Florian Mendel, Bart Preneel, and Christian Rechberger. Collisions and other non-random properties for step-reduced SHA-256. In Roberto Maria Avanzi, Liam Keliher, and Francesco Sica, editors, *Selected Areas in Cryptography*, volume 5381 of *Lecture Notes in Computer Science*, pages 276–293. Springer, 2008.

[37] NIST. First cryptographic hash workshop and second cryptographic hash workshop. National Institute of Standards and Technology, 2005/2006.

[38] *Announcing Request for Candidate Algorithm Nominations for a New Cryptographic Hash Algorithm (SHA-3) Family*. NIST, 2007. http:// csrc.nist.gov/groups/ST/hash/index.html.

[39] NIST. SHA-3 first round candidates. National Institute of Standards and Technology, 2008.

[40] NIST. SHA-3 second round candidates. National Institute of Standards and Technology, 2009.

[41] NIST. National Institute of Standards and Technology, 2010.

[42] Alfred J. Menezes, Paul C. van Oorschot, and Scott A. Vanstone. *Handbook of Applied Cryptography*. CRC Press, 2001.

[43] C. P. Pfleeger and S. L. Pfleeger. *Security in Computing*. Prentice Hall PTR, 2003.

[44] K. H. Rosen. *Handbook of Discrete and Combinatorial Mathematics*. CRC Press, 2000.

[45] International Association for Cryptologic Research. http://www.iacr. org/.

5

Quantum Cryptography

D. R. Hjelme

Department of Electronics and Telecommunications, NTNU

L. Lydersen

Department of Electronics and Telecommunications, NTNU, and Kjeller University Graduate Center

V. Makarov

Department of Electronics and Telecommunications, NTNU, and Kjeller University Graduate Center

CONTENTS

5.1   Introduction

When information is transmitted in microscopic systems, such as single photons (single light particles) or atoms, its information carriers obey quantum rather than classical physics. This offers many new possibilities for information processing, since it is possible to invent novel information processes prevented by classical physics.

Quantum cryptography is the most mature technology in the new field of quantum information processing. Unlike cryptographic techniques where the security is based on unproved mathematical assumptions, the security of quantum cryptography is based on the laws of physics. Today, it is developed with an eye toward a future in which cracking of classical public-key ciphers might become practically feasible. For example, a quantum computer might one day be able to crack today's codes. For instance, the security of RSA public-key cryptography (Chapter 3) rests on the widely-believed assumption that the factorization problem is computationally hard. Although no efficient factorization algorithm is publicly known, it has *not* been proved that one does not exist. Shor's algorithm for a *quantum computer* already allows efficient factorization; however, it remains an open question if and when a scalable quantum computer is built. Furthermore, once a classical encryption is broken, the crack can be applied to today's secrets *retroactively*. This is uncomfortable for many types of secret information whose value persists for decades: government and military communication, commercial secrets, as well as certain personal information such as financial and medical records.

The one time pad [1] remains unassailable even by such future techniques. The weakness of the one time pad is that a secret, random, symmetric key as long as the message it is intended to encrypt must be securely distributed to the intended receiver of the message. Furthermore, the key can only be used once. This is infeasible using only classical physics. Quantum cryptography solves this key distribution problem by exploiting how single quantum particles behave.

The working principles of quantum cryptography can simply be explained by considering information transmission using single photons. A single photon can represent a quantum bit, a so-called *qubit*. To determine the qubit value one must measure the representing property of the photon (for example, polarization). According to quantum physics, such a measurement will inevitably alter the same property. This is disastrous for anyone trying to eavesdrop on the transmission, since the sender and receiver can easily detect the changes caused by the measurement. Since the security can only be determined after a transmission, this idea cannot be used to send the secret message itself. However, it can be used to transmit a secret, random, symmetric key for one time pad cryptography. If the transmission is intercepted, the sender and receiver will detect the eavesdropping attempt,[2] the key can be discarded and the sender can transmit another key until a secure key is received.

In spite of the simple principles behind quantum cryptography, the idea was first conceived as late as 1970 in an unpublished manuscript written by Stephen Wiesner . The subject received very little attention until its resurrection by a classic paper published by Charles Bennett and Gilles Brassard in 1984. Currently, the technology required for quantum cryptography is available for real-world system implementations.

The objective of this chapter is to present the working principle of quantum cryptography and to give examples of quantum cryptography protocols and implementations using technology available today. Throughout the chapter, we minimize the use of quantum physics formalism and no previous knowledge of quantum physics is required. References are provided for the interested reader who craves for more details. A good starting point is the excellent review by Gisin *et al.* [2]; also the original paper [3] explains the quantum cryptography protocol very well.

---

[1] It has been proved that a secure cipher needs to use the amount of secret key at least as large as the length of the message [1]. The one time pad (Section 3.2) is one such cipher.

[2] No such possibility exists if the key is exchanged using classical physics because classical bits can be read, and hence copied without the risk of destroying the original bit value.

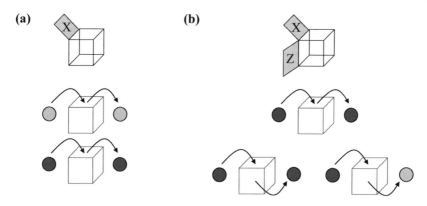

FIGURE 5.1
Classical versus quantum bit. (a) Classical bit: If we put the ball in a classical box, the color of the ball that pops out is the same as the color we put in. (b) Qubit: If we put the ball in a quantum box and open the wrong door, the color of the ball that comes out is random.

5.2   Quantum Bit

All information can be reduced to elementary units, which we call bits. Each bit is a *yes* or *no* that can be represented by the number 0 or the number 1. However, as we will see, reading and writing this information to a qubit is something quite different from reading and writing this information to a classical bit.

We can think of a (qu)bit as a box, where we can store one of the two bit values by putting a ball with one out of two colors into the box as illustrated in Figure 5.1. To read the bit value of the box, we simply open the box and register the color of the ball inside. For the classical bit, the color of the ball inside is always the same as the color of the ball stored in the box in the first place. However, this is not necessarily the case for qubits.[3] In the quantum formalism, the two different doors of the box represent two different ways of measuring the qubit value. To read the correct bit information, we need to know which door was used when the qubit was stored, and use the same door. If we open the wrong door, the ball inside will have a random color, and thus the information stored in the qubit will change to a random bit value. This also means that the stored information is destroyed.

One realization of the qubit is a polarized photon . One way of determining the polarization of the photon is to send it through a polarizing beamsplitter, and measure at which output of the polarizing beamsplitter the photon is

---

[3]We have borrowed this way of visualizing a qubit from John Preskill [4].

(a) **(b)** **(c)**

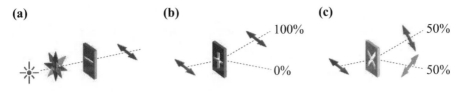

FIGURE 5.2

Qubit as a polarized photon. (a) A photon source is followed by a linear polar-
izer to generate a qubit with the desired polarization, in this case a horizontally
polarized photon. (b) When a horizontally polarized photon passes through a
horizontally-vertically oriented polarizing beamsplitter, it is always found at
the exit of the beamsplitter corresponding to the horizontal polarization. (c)
When a horizontally polarized photon passes through a diagonally oriented
beamsplitter, the photon has 50% probability to be found at each exit (but
the photon will only be detected at one of the exits!). Furthermore, the photon will have a corresponding diagonal polarization afterwards. Therefore, the
measurement has changed the state of the photon.

found.[4] However, since the polarizing beamsplitter only separates between
orthogonal polarizations, we cannot orient the polarizing beamsplitter at two
angles at the same time. Thus, we cannot read the qubit value unless we have
additional information. For example, if we know that the polarization is either
horizontal or vertical in a defined reference coordinate system,[5] we can read
the qubit value by orienting the beamsplitter to the axis of the coordinate
system. If we find a photon at one output of the beamsplitter, we know that
the photon polarization was horizontal; if we find it at another output, the
polarization was vertical (see Figure 5.2). That is, we need to know a priori
which coordinate system is used in preparing the qubit to read it correctly. If
we use another orientation of the beamsplitter, the result of the measurement
will be random just like when opening the wrong door of the quantum box in
Figure 5.1. Note that once the photon has been detected in one of the outputs
after the beamsplitter, the photon actually assumes the output polarization,
with no trace of its original polarization left – this is how the nature works at
the quantum level.

---

[4]A polarizing beamsplitter is a device that separates orthogonal linear polarizations of
incoming light into two directions. An example is *Wollaston prism.*

[5]In quantum physics, the orientation of the beamsplitter is called the *basis.*

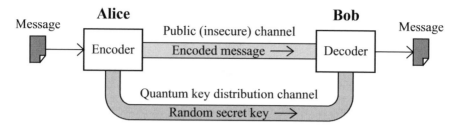

FIGURE 5.3
Using quantum key distribution in a symmetric encryption scheme. The first step is distribution of a secret key between Alice and Bob. Then, the key can be used by a symmetric cipher to encode and decode transmitted information.

5.3   Quantum Copying

To copy a qubit, we need to read the bit value; that is, we need to open the quantum box. However, there is no way of knowing which door was used to store the bit value of the qubit. If we simply guess one of the doors, we may damage the information stored in the qubit. Thus generally, since quantum bits cannot be perfectly read, quantum bits cannot be perfectly copied either.[6]

Usually, the ability to copy information is considered to be very useful. But, in secure communication, this would be disastrous since the eavesdropper could listen to the communication and keep a copy of the message. However, qubits cannot be copied. This noncopying property of quantum information can be exploited for secure communication. Therefore qubits can be used to distribute a key from sender to recipient without the possibility for the eavesdropper to obtain a copy surreptitiously.

5.4   Quantum Key Distribution

Quantum cryptography is not used directly to transmit the secret information, but is rather used to distribute a random secret key, see Figure 5.3. Once the key has successfully been transmitted, it can be used in a classical symmetric cipher (such as the one time pad described in Section 3.2 or AES described in Section 2.2.2) to encrypt and decrypt information. Let's consider the quantum key distribution protocol.

---

[6]For a strict quantum-mechanical proof of this fact, see [5]. The proof is very short.

5.4.1 The BB84 Protocol

To explain the protocol, let us call the sender Alice, and the receiver Bob. Assume that Alice generates a random sequence of bits, codes them in qubits randomly using door X or door Z of the quantum box, and sends the qubits over a quantum channel to Bob. Bob does not know which doors Alice used, and therefore he randomly picks doors. The result is that Bob opens the right door only half the time. In those cases, he reads the right information. Bob's bits are called the raw key at this stage. After Bob has opened all the quantum boxes, both he and Alice publicly announce which doors were used to store and measure the qubits values. They then keep only the qubit values from the boxes where they happened to use the same doors. This random sequence of bits now shared by Alice and Bob is called the sifted key, and is about half as long as the original raw key.

What happens if the eavesdropper Eve tries to open some of the quantum boxes during the transmission? If Eve by chance opens the right door, she can copy the information and send it to Bob. However, half of the time she will open the wrong door and might change the value of the qubit. If Alice and Bob conduct a test and compare a small portion of their key, they can make sure that Bob received what Alice sent. If Alice's and Bob's portion of the key matches, they can be confident that Eve did not open any boxes. On the other hand, if their keys do not agree, they know that Eve tried to measure the key.

What we have just described is the quantum key distribution protocol BB84, first presented in 1984 by Bennett and Brassard. Given that Alice and Bob can only measure the fraction of errors in the key, often called the *quantum bit error rate*, the protocol either delivers a provably secure key, or informs Alice and Bob that the key distribution failed.

5.4.2 The BB84 Protocol Using Polarized Light

The BB84 protocol can be implemented using polarized single photons as illustrated in Figure 5.4. Alice codes the qubit using horizontal (bit value 0) and vertical (bit value 1) polarization, or she codes the qubit using $-45°$ (bit value 0) or $+45°$ (bit value 1) polarization.[7] To receive the qubits, Bob uses two interchangeable polarizing beamsplitters and two photon detectors[8] after the beamsplitter . One polarizing beamsplitter allows Bob to distinguish between the horizontal and vertical polarizations, and the other polarizing beamsplitter allows Bob to distinguish between the $-45°$ and $+45°$ polarizations. If Bob uses a polarizing beamsplitter compatible with the polarization choice of Alice, he will read the state of polarization correctly; that is, he opened the right

---

[7]These two ways of doing the coding represent the two doors of the quantum boxes described earlier.

[8]A photon detector is a device that gives a signal ('click') when a photon arrives at the device.

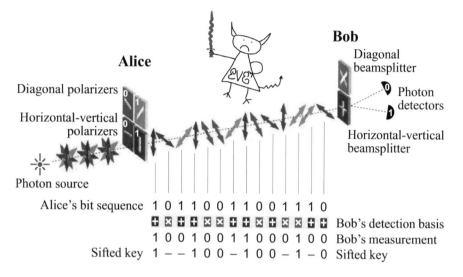

FIGURE 5.4
BB84 protocol using polarized light (reprinted from [6]).

door. If Bob uses a polarizing beamsplitter incompatible with the polarization choice of Alice he will not be able to get any information about the state of polarization; that is, he opened the wrong door.

After receiving enough photons, constituting the raw key, Bob announces over a public classical communication channel (for example, over an internet connection) the sequence of polarizing beamsplitters he used but, importantly, not the result of the measurement. Alice compares this sequence to the sequence of bases (polarization choice) she used and tells Bob on which occasions he used the right beamsplitter[9] but, importantly, not the polarization she sent. For these bits, constituting the sifted key, Alice and Bob know that they have the same bit values provided that an eavesdropper did not perturb the transmission.

To assess the security of the transmission, Alice and Bob select a random subset of the sifted key and compare it over the public channel. If the transmission were intercepted or perturbed, the correlation between their bit values will be reduced, thus increasing the quantum bit error rate. All eavesdropping strategies perturb the system in some way. Therefore, if Alice and Bob do not measure any discrepancy in the subset of the key, they can be confident that the transmission was not intercepted and they can use the remaining part of the key for encryption.

---

[9]Effectively publicly announcing which door of the box was used to store each qubit.

5.5   Practical Quantum Cryptography

Any implementation of quantum key distribution uses technology available today, meaning that the system components such as photon sources, transmission channel, polarizing beamsplitters, and photon detectors are imperfect. This fact has several important implications.

5.5.1   Loss of Photons

One imperfection common to all components is that photons sometimes get lost. In a practical system, the majority of the photons exiting Alice will get absorbed in the transmission channel, and those that reach the detector will often fail to cause a click. In practice only the photons that have registered as clicks in Bob's detectors contribute bits to the raw key.

5.5.2   Error Correction and Privacy Amplification

Another implication of imperfections is that the qubits are prepared and detected not exactly in the basis as described by theory. Technological imperfections will lead to errors in the sifted key, errors that cannot be distinguished from errors resulting from any eavesdropping attempts. Realistic error rates with today's technology are in the order of a few percent. This quantum bit error rate is often dominated by false detection signals from the photon detectors, so-called *dark counts*. [10]

Alice and Bob cannot be sure whether the errors in the sifted key resulted from device imperfections or from eavesdropping. They have to assume the worst, and assume all errors were due to eavesdropping. At this point in the protocol, Alice and Bob share classical information with high but not 100% correlation, and assume that the third party Eve has partial knowledge of this information. This problem can be solved by classical information theory, which has methods of distilling a shorter, error-free key of which Eve has no knowledge about.

First, Alice and Bob need to apply classical error correction techniques to obtain identical keys. [11] Eve still knows some information about this key (actually she knows even more than before, because Alice and Bob have had to reveal more information while communicating publicly during the error correction). The last step in the quantum cryptography protocol therefore is a *privacy amplification* procedure that shrinks the key and reduces the amount

---

[10] Dark counts are clicks in detector without any photons present, and can thus be observed at the detector output in the dark.

[11] Very high raw error rate of a few percent, while typical for quantum cryptography, usually does not occur in classical telecommunication. Therefore, special error correction algorithms have been developed for quantum cryptography.

of information Eve may know about it. Alice and Bob do privacy amplification by applying a randomly chosen hash function of *universal₂*-class to the error-corrected key.[12] As long as Bob has more information about Alice's sifted key than Eve, privacy amplification will produce a shorter final key about which Eve's information is arbitrarily small. To give a feel for the numbers, with realistic quantum bit error rate of 4%, assumed to be dominated by eavesdropping, 2000 bit can be distilled down to 754 secret bit about which Eve's information is negligible (less than 10^{-6} bit). With quantum bit error rate of 8% we can distill 105 secret bit from the original 2000 bit [3].

The resulting workflow of a general quantum key distribution algorithm is illustrated in Figure 5.5.

5.5.3 Security Proofs

The intuition as to why quantum key distribution provides perfectly secret key is quite straightforward. However, the details of the proofs are very involved [8]. If one assumes that Eve can only interact with one qubit at a time,[13] and that Alice and Bob are using a perfect implementation of the protocol, it has been proved that Eve will never know as much as Bob provided that the quantum bit error rate is less than 14.65%. If Eve has unlimited power and can coherently attack an unlimited number of qubits,[14] that is, she can do everything allowed by the known laws of physics, it has been proved that a quantum bit error rate less than 11% is required for secure communication. As long as the error rate is below this threshold, the security proof provides an equation that can be used to compute the required amount of privacy amplification (Figure 5.5).

The security has been proved strictly for certain idealized models of equipment. However, most of the current discussion is whether imperfections in real hardware (not yet accounted by the proofs) may leave loopholes, and how to close these loopholes [9].

5.5.4 Authentication

One problem remains: How can Alice and Bob be sure they really talk to each other on the public channel and not to Eve, when they produce the secret key? Eve could be in the middle between Alice and Bob, representing herself as Bob to Alice and as Alice to Bob. The prevention of this is known and requires that Alice and Bob start from an initial short common secret (a few hundred bit), so as to be able to recognize each other during their first run

---

[12]These hash functions and this application are different from those described in Chapter 4. While the security of cryptographic hash functions in Chapter 4 in not proved, here the security of the privacy amplification procedure [7] is *unconditional;* that is, strictly proved against an adversary who possesses unlimited computing power.

[13]This is a so-called individual attack.

[14]This is a so-called coherent attack.

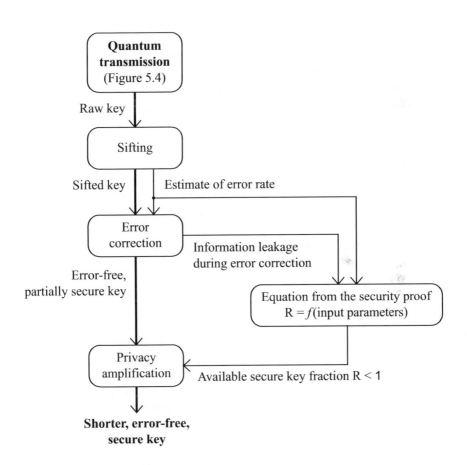

FIGURE 5.5
Classical post-processing in quantum key distribution. Alice and Bob start with the raw photon detection data, communicate over an authenticated classical channel while performing sifting, error correction and privacy amplification procedures, and arrive at a secret shared key about which Eve has negligible information.

of the protocol.[15] After the first successful key distribution, they can use a part of the secret key they produce to authenticate in future runs. It has been proved that quantum key distribution provides much more secret key than it consumes in authentication.[16] In this sense, quantum key distribution is a quantum secret-growing protocol.

The need for initial authentication is intrinsic and universal to all flavors of cryptography: How else can you verify that you are talking to the intended party and not to Eve? The initial trusted key and/or biometric authentication (by, for example, verifying a pen signature, talking to a known person over phone or being physically present during the transaction) is found in some form in all cryptographic protocols.

5.6   Technology

Essentially two technologies make quantum cryptography possible: single photon sources and single photon detectors. In addition, a transmission channel for the single photon states, so-called quantum channel, is needed. The rest of the system is realized using fairly standard telecommunication and electronic hardware.

5.6.1   Single Photon Sources

Single photon sources are difficult to realize. Therefore, most systems today rely on faint laser pulses. Conventional laser pulses, for example, from a semiconductor laser, are attenuated such that there is on the average less than one photon per pulse. The problem with this approach is that there is a significant probability that there are two or more photons per pulse, unless the average photon number is far below one. The number of photons in the pulse follows Poisson statistics, which for instance means that in a pulse of average photon number 0.1, there is a 0.9048 probability to find no photons, 0.0905 probability to find one photon, and 0.0047 probability to find two or more photons. If Alice emits pulses containing more than one photon, Eve can take and store

---

[15]Unconditionally secure authentication is employed, using hash functions of 'almost' *universal*$_2$-class [10]. The secret key is used to pick a function from the set of hash functions, then that function is applied to the message to compute a shorter authentication tag. The message and authentication tag are sent to the other communicating party. The latter computes an authentication tag on the received message using her copy of the secret key, and compares it with the received tag. If the tags are the same, this guarantees that the messages are the same with a high probability. The use of one time pad to pick the hash function guarantees against attacks on this authentication scheme.

[16]While perfect encryption, for example, the one time pad needs m secret bits to encrypt an m-bit message, perfect authentication only needs in the order of $\log(m)$ secret bits to authenticate an m-bit message.

one of the photons in the pulse until the basis is announced. Then she may perform a perfect measurement in this basis, learning the bit value of the qubit sent to Bob. Therefore, the presence of multiphoton pulses decreases the secret key rate. The fraction of multiphoton pulses relative to single-photon pulses can be reduced by decreasing the average photon number. However, when the average photon number is small it means that most bit slots are empty, also resulting in lower bit rate. In principle, the latter could be compensated for by increasing the pulse rate. However, another drawback remains, as the dark counts (false detection events) in the single photon detectors are significant. The result is that the signal-to-noise ratio decreases, raising the quantum bit error rate, as the average photon number decreases.

The ideal photon source is a device that emits single photons on demand.[17] Although progress is reported, practical devices are not yet available [11].

Nevertheless, practical operation over tens of kilometers has been achieved using faint laser pulse sources. Also, there are advanced protocols[18] that allow secure operation over longer than 100 km distance with the faint laser pulse source.

5.6.2 Single Photon Detectors

Single photon detection can be realized in a number of ways, for example, using photomultipliers, avalanche photodiodes, as well as several types of more exotic superconducting devices that have to be cryogenically cooled below 4 K.[19] Today, the best and in fact the only practical choice for quantum cryptography is the avalanche photodiode [14]. An avalanche photodiode is a semiconductor component, and to detect single photons it is operated under a large voltage.[20] If a single photon is absorbed by the semiconductor, it excites a single electron. The high electric field in the semiconductor ensures that this initial electron collides with the lattice and excites more electrons, thus being amplified into an avalanche of electrons (several thousands). This avalanche is large enough to be detected as a current pulse by an external circuit. Unfortunately, an avalanche can also occur without a photon, initiated by thermal excitation, tunneling, or emission of trapped carriers. The latter happens when electrons from a previous avalanche get stuck in defects of the semiconductor lattice, then slowly released. This emission of trapped carriers limits the practical count rate. This is a serious limitation in the current systems using faint laser pulses, where a high pulse rate is desirable in order to achieve acceptable bit rates.

---

[17]Such a source is often called a *photon gun*.

[18]For example, the decoy-state protocol [12].

[19]For a wide review of photon detection techniques, see [13].

[20]The photodiode is reverse biased above its breakdown voltage.

5.6.3 Quantum Channel

Alice and Bob must be connected by a quantum channel. This channel must be such that the qubit is protected from environmental noise. Standard single-mode optical fiber used for data and telecommunication is an almost ideal channel for single photon states (qubits). All optical fibers have transmission losses limiting the number of qubits arriving at the detector. This has direct impact on the key exchange rate, as the raw key rate is directly proportional to the photon transmission probability of the link. Modern telecommunication fibers have transmission losses of about 2 dB/km, 0.35 dB/km, and 0.2 dB/km in the commonly used communication wavelength windows of 800 nm, 1300 nm, and 1550 nm, respectively. At 1550 nm this means that at least 50% of the photons are lost at 15 km, or 99% of the photons are lost at 100 km. The longest successful quantum key distribution reported in laboratory conditions to date is 250 km, at a very slow rate of 15 secret bit/s indeed [15]. Today's commercial systems are limited to 50–100 km.

All fibers are subject to environmental fluctuations, such as a change in temperature. This perturbs a polarization state, and therefore changes the qubit values. Thus, the error rate is increased by the imperfect channel. The global effect of this polarization state perturbation is a transformation between the fiber input and fiber output. If this transformation is stable, Alice and Bob can compensate for it by using a polarization controller to align their systems by defining, for example, the vertical and diagonal polarization direction. If the transformation varies slowly, one can use an active feedback system to maintain alignment over time. Smart solutions are possible: early commercial systems used a so-called plug and play optical scheme that canceled polarization perturbation without a need for active control [16].

As an alternative to fiber, a line-of-sight path via atmosphere can be used as the quantum channel. Alice and Bob use small telescopes pointed at each other to transmit photons. Availability and quality of such a link is obviously affected by weather conditions. However, air neither perturbs polarization nor has a high loss. The longest transmission has been achieved over 144 km between hilltops on the Canary Islands [17]; however, links of 10–30 km may be more practical [18]. The success of these ground experiments suggests a possibility of distributing a secret key between a ground station and a satellite. A low-orbit satellite can thus provide a global key distribution network by successively establishing key distribution links with places under its flight path.[21]

5.6.4 Random Number Generator

The key used for the one time pad must be perfectly random. As computers are deterministic systems, they cannot be used to create random numbers for cryptographic systems. Therefore, the random numbers must be created by a

---

[21]This requires the satellite to operate as a trusted node, as discussed later in this chapter.

truly random physical process. One example is a single photon sent through a beamsplitter: the photon is found in one of the two exits of the beamsplitter. Which exit it is found at, is random according to quantum mechanics. There are certainly many other processes that can be used. While physical random number generators with bit rates of a few Mbit/s are employed in current commercial quantum key distribution systems, construction of high bit rate true random number generators is still at an experimental stage.

5.7 Applications

5.7.1 Commercial Application of Quantum Cryptography

Although quantum cryptography is quite technologically mature, and commercially it currently enjoys a tiny niche market, perspectives of wider adoption are unclear. On the one hand, classical cryptographic systems based on assumptions on computational complexity are very good and convenient: well-developed, cheap, can work at high bit rates over unlimited distance. As we have discussed in the Introduction of this chapter, their security is not guaranteed against future advances in cryptanalysis (and strictly speaking, not even guaranteed today), but their convenience is almost unbeatable. Should any of them fall, this would not be the first historical example when the ease of use was preferred over stricter security [19].

On the other hand, a quantum key distribution link is limited by distance and bit rate, and is currently relatively expensive to set up. In fact, sending a person (trusted courier) carrying a hard disk filled with random numbers would provide a larger key supply than most quantum key distribution links could deliver over their operation lifetime. Note that the quantum key distribution link also needs a short key for the initial authentication, so the trusted courier is involved anyway. However, the ability to grow key limitlessly from the short authentication key makes quantum cryptography scale well in a network of many users, while in the hard disk distribution scenario the required storage capacity would quickly become unrealistic [20].

5.7.2 Commercial Systems with Dual Key Agreement

In today's commercial systems, quantum key distribution is used as an extra security layer on top of classical key distribution and encryption, see Figure 5.6. Keys obtained from quantum key distribution are combined with keys sent using public key cryptography, by encrypting one key using the other key as one time pad (exclusive-OR binary function). The resulting combined key is at least as secure as the stronger of the two original keys. Thus, to eavesdrop the combined key, an attacker would have to crack both public-key cryptography *and* quantum key distribution. This combined key is changed several times a second, and used in a high-throughput symmetric cipher to encrypt a wideband network link.

Although any symmetric cipher using key shorter than the encrypted message is not unconditionally secure, this architecture is dictated by the ease of integration into existing networks. Customers are used to having classical cryptography that can encrypt the entire gigabit network link. Nevertheless, it is argued that the security of the AES symmetric cipher improves when the key is changed frequently, and thus less ciphertext is available for cryptanalysis.

The system has an option to additionally provide one time pad encryption to the users. In the commercially available units, the key generation speed and thus one time pad average bandwidth is no faster than a few kbit/s. However, laboratory prototypes have been demonstrated with up to 1 Mbit/s over 50 km fiber [21].

5.7.3   Quantum Key Distribution Networks

To increase the number of users and overcome the link distance limitation, two types of networks are possible: with trusted nodes, and with untrusted nodes. The trusted-node network consists of point-to-point quantum key distribution links between nodes. When two users want to establish a shared key, they find a path through intermediate nodes, then one user sends his key to the other user through a chain of one time pad encryptions using keys generated in each point-to-point link along the path. This type of network has been demonstrated in several metropolitan areas [22, 23].

The untrusted-node network can use optical switches at the nodes to create an uninterrupted optical channel between end users. This is realistic with today's technology, but the optical switches do not increase transmission distance, and can thus only be used in a geographically compact network. An alternative idea is to use so-called *quantum repeaters* at the untrusted nodes, which in theory can increase the distance far beyond the 250 km limit. However, quantum repeaters remain a future technology. The untrusted-node network configuration can realize the full potential of quantum cryptography, and perhaps provide a decisive advantage over using trusted couriers and other key distribution methods. For example, each user can get and store only initial authentication keys for every other network user, then grow more key material with any user as needed.

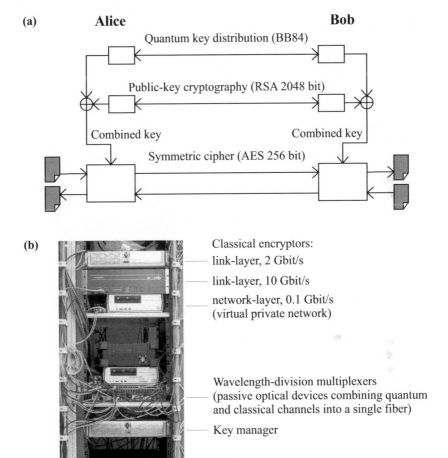

FIGURE 5.6

Commercial quantum cryptography vintage 2010. (a) Dual key agreement scheme. Two secret keys are distributed independently using quantum cryptography and public-key cryptography, then added modulo 2 (X-ored) together. The resulting key is used in a symmetric cipher to encrypt network traffic. (b) Network node with quantum key distribution equipment, in a standard 19 inch wide server rack. Quantum keys are generated between this node and two other remote nodes, using ID Quantique Vectis units, then passed to classical equipment that encrypts all network traffic with those remote nodes. This node was a part of SwissQuantum testbed network in Geneva, and operated continuously for more than a year, see http://swissquantum.idquantique.com/.

5.8   Summary

The feasibility of quantum cryptography has now been demonstrated over distances up to 250 km, and in key distribution networks. Although the systems still suffer from low key transmission rates, they do provide means for secure communication if the public-key systems used today are not trusted. But foremost, quantum cryptography is developed today with an eye toward a future in which cracking classical public-key ciphers might become practically feasible. For example, a quantum computer might one day be able to crack today's codes. Quantum cryptography is also an excellent example of the intimate interplay between fundamental and applied research.

5.9   Further Reading and Web Sites

The web site http://www.iet.ntnu.no/groups/optics/qcr/ of our Quantum Hacking group presents how industrial implementations of the quantum key distribution system can be broken. The web site http://pqcrypto.org/ investigates what will happen to cryptology when the first working quantum computer has been built.

Bibliography

[1] C. E. Shannon. Communication theory of secrecy systems. *Bell System Technical Journal*, 28:656–715, 1949.

[2] N. Gisin, G. Ribordy, W. Tittel, and H. Zbinden. Quantum cryptography. *Rev. Mod. Phys.*, 74:145, 2002.

[3] C. H. Bennett, F. Bessette, L. Salvail, G. Brassard, and J. Smolin. Experimental quantum cryptography. *Journal of Cryptology*, 5:3–28, 1992.

[4] J. Preskill. Making weirdness work: Quantum information and computation. In *IEEE Aerospace Conference 1998 proceedings*, volume 1, pages 37–46, 1998.

[5] W. K. Wootters and W. H. Zurek. A single quantum cannot be cloned. *Nature*, 299:802, 1982.

[6] W. Tittel, G. Ribordy, and N. Gisin. Quantum cryptography. *Physics World*, 11:41–45, March 1998.

[7] C. H. Bennett, G. Brassard, C. Crépeau, and U. M. Maurer. Generalized privacy amplification. *IEEE Transactions on Information Theory*, 41:1915–1923, 1995.

[8] V. Scarani, H. Bechmann-Pasquinucci, N. J. Cerf, M. Dušek, N. Lütkenhaus, and M. Peev. The security of practical quantum key distribution. *Reviews of Modern Physics*, 81:1301–1350, 2009.

[9] I. Gerhardt, Q. Liu, A. Lamas-Linares, J. Skaar, C. Kurtsiefer, and V. Makarov. Full-field implementation of a perfect eavesdropper on a quantum cryptography system. *Nature Communications*, 2:349, 2011.

[10] M. N. Wegman and J. L. Carter. New hash functions and their use in authentication and set equality. *Journal of Computer and System Sciences*, 22:265–279, 1981.

[11] A. J. Shields. Semiconductor quantum light sources. *Nature Photonics*, 1:215–223, 2007.

[12] W.-Y. Hwang. Quantum key distribution with high loss: Toward global secure communication. *Physical Review Letters*, 91:057901, 2003.

[13] R. H. Hadfield. Single-photon detectors for optical quantum information applications. *Nature Photonics*, 3:696–705, 2009.

[14] S. Cova, M. Ghioni, A. Lotito, I. Rech, and F. Zappa. Evolution and prospects for single-photon avalanche diodes and quenching circuits. *Journal of Modern Optics*, 51:1267–1288, 2004.

[15] D. Stucki, N. Walenta, F. Vannel, R. T. Thew, N. Gisin, H. Zbinden, S. Gray, C. R. Towery, and S. Ten. High rate, long-distance quantum key distribution over 250 km of ultra low loss fibres. *New Journal of Physics*, 11:075003, 2009.

[16] A. Muller, T. Herzog, B. Huttner, W. Tittel, H. Zbinden, and N. Gisin. "Plug and play" systems for quantum cryptography. *Applied Physics Letters*, 70:793–795, 1997.

[17] R. Ursin, F. Tiefenbacher, T. Schmitt-Manderbach, H. Weier, T. Scheidl, M. Lindenthal, B. Blauensteiner, T. Jennewein, J. Perdigues, P. Trojek, B. Ömer, M. Fürst, M. Meyenburg, J. Rarity, Z. Sodnik, C. Barbieri, H. Weinfurter, and A. Zeilinger. Entanglement-based quantum communication over 144 km. *Nature Physics*, 3:481–486, 2007.

[18] C. Kurtsiefer, P. Zarda, M. Halder, H. Weinfurter, P. M. Gorman, P. R. Tapster, and J. G. Rarity. A step towards global key distribution. *Nature*, 419:450, 2002.

[19] S. Singh. *The Code Book*. Random House, New York, 1999.

[20] L. Lydersen. *Practical security of quantum cryptography.* PhD thesis, Norwegian University of Science and Technology, 2011.

[21] A. R. Dixon, Z. L. Yuan, J. F. Dynes, A. W. Sharpe, and A. J. Shields. Continuous operation of high bit rate quantum key distribution. *Applied Physics Letters*, 96:161102, 2010.

[22] N. Horiuchi. View from... UQCC 2010: Quantum secure video. *Nature Photonics*, 5:10–11, 2011.

[23] M. Peev, C. Pacher, R. Alléaume, C. Barreiro, J. Bouda, W. Boxleitner, T. Debuisschert, E. Diamanti, M. Dianati, J. F. Dynes, S. Fasel, S. Fossier, M. Fürst, J.-D. Gautier, O. Gay, N. Gisin, P. Grangier, A. Happe, Y. Hasani, M. Hentschel, H. Hübel, G. Humer, T. Länger, M. Legré, R. Lieger, J. Lodewyck, T. Lorünser, N Lütkenhaus, A. Marhold, T. Matyus, O. Maurhart, L. Monat, S. Nauerth, J.-B. Page, A. Poppe, E. Querasser, G. Ribordy, S. Robyr, L. Salvail, A. W. Sharpe, A. J. Shields, D. Stucki, M. Suda, C. Tamas, T. Themel, R. T. Thew, Y. Thoma, A. Treiber, P. Trinkler, R. Tualle-Brouri, F. Vannel, N. Walenta, H. Weier, H. Weinfurter, I. Wimberger, Z. L. Yuan, H. Zbinden, and A. Zeilinger. The SECOQC quantum key distribution network in Vienna. *New Journal of Physics*, 11:075001, 2009.

6

Cryptographic Protocols

S. F. Mjølsnes

Department of Telematics, NTNU

CONTENTS

6.1　The Origins

> *Security protocols are three-line programs*
> *that people still manage to get wrong. — Roger Needham*

The classical model of a crypto-system represents the simplest protocol possible. It is a one-way transmission that involves two parties, the sender and the recipient. The channel is a common object for which the sender can write to and the recipient can read from. The goal is to communicate while keeping the information on the channel incomprehensible to outside parties that have read access to the channel. A common secret cryptographic key makes it possible to carry this through. Whoever acquires the key will be capable of computing the cleartext from the ciphertext. The cryptographic function must at least be designed to withstand a ciphertext-only attack. If side-information might be available then the function must be analyzed with respect to a known-plaintext attack .

Another class of the very early cryptographic protocols is the *Identification-friend-or-foe systems* (IFF) . The development of the IFF protocols started in 1952 by Feistel's group at The Air Force Research Center, Cambridge, USA. The problem was to distinguish between friendly and hostile aircrafts that approach the base. The fire control radar classifies the aircraft by challenging it and, subsequently, checking the reply. A constant reply will not be of any help, because it is easily picked up by the enemy and played back whenever they themselves are challenged. The solution adopted by Feistel's group, and now used in both military and civil applications, varies the exchange cryptographically each time the authentication takes place. The challenge is selected at random, and the response is correct only when properly encrypted. Ideally, the challenges are never repeated, and hence previously recorded responses will not be of any direct help. The cryptographic function must be designed to withstand a known-plaintext attack or even a chosen-plaintext attack, because the interaction is public and anybody may send a challenge. If the secret key is revealed, then the authentication fails.

Conversely, consider the function to be public knowledge and restrict it to be a one-way function . A one-way function is characterized by the property that if image values are given then this fact does not imply any feasible way of finding the corresponding inverse image values. A computational barrier exists from output to input, and only in that order. Roger Needham of Cambridge University, UK, is reported to be the first to have applied this concept in operating systems. By assuming some one-way function and a constant secret response, Needham solved the problem of protecting computer passwords. Instead of having to guard the secret password tables against unauthorized reading and writing, he introduced a table of derived values: the images of the passwords under a one-way function. It is the input that is secret, not a key nor a function, in this protocol.

Public-key cryptography inventor Whitfield Diffie describes how he realized that one could combine the two cryptographic protocols and solve both problems at the same time [1]. The challenge should be chosen at random as before, but now only the challenged party can compute the correct response. However, the challenger can check the response efficiently. Basic to the invention is the splitting of capabilities. The challenged party can compute something *nobody* else can. If the input is public, the result is called a signature . If the input is ciphertext, the result is cleartext. The difference is clear: in the challenge-response scheme the function must be kept secret by keeping some secret key. The function could be public in the password scheme. In the public-key scheme, "half" the function is public. The generalization now follows and becomes the trapdoor one-way function.

Public key cryptography introduces the trapdoor one-way function and differentiates the capabilities of the sender and the recipients. The number of composition possibilities is magnified by this. Now there are two functions or operations special to every participant in the protocol. This partly solves the key-distribution problem of multiparty communications. The secrecy of the encryption key is no longer needed, but the authenticity of the key remains a concern.

Zero-knowledge interactive proof protocols are a very spectacular invention of cryptographic research, where the goal is for a prover to convince a verifier of the possession of some knowledge without revealing additional information. The conviction of the verifier cannot be transferred to someone else by mathematical means. These protocols are based on one-way functions and resemble a form of challenge-response authentication where both respondent and challenger are allowed to randomize, but only the respondent has the trapdoor. It is not surprising that a major application of the zero-knowledge protocols is proofs of identity. The traces of the origins of public key protocols are evident.

6.2   Information Policies

Consider some set of networked computers and the following game. Before allowing the computers to start interacting, specific pieces of information, such as program and data files, are assigned to each computer. Thereafter, the interactions start and cause information to be spread among the machines and disseminated on the network. The aim is now for each computer *to control* as much as possible of the information flow according to preset conditions. The conditions, or information policy, may range from concerns about not revealing specific information to concerns about ensuring that the correct and complete information is timely revealed and preserved in the computer domains determined. The following list indicates a classification of some concerns that may arise:

- Release of information:

    - Maintain private information.
    - Prove the possession of secrets without releasing more information.
    - Exclusive sharing of information.
    - Gradual release of information.
    - Oblivious release of information.
    - Exchange of secret information.
    - Anonymity of sender and recipient.

- Preservation of information:

    - Maintain correct and complete information among two or more parties.
    - Correct sender(s) and recipient(s).
    - Correct time and complete sequence of events.

We can recognize these concerns in most application areas of networked computers and communications. For example, e-commerce transactions, online auctions, and financial trading systems require accountability and fairness in the information exchange. Health registries require availability and privacy.

The problem of how to let voters cast their ballots over the Internet contains a rich set of security challenges and is perhaps the single most difficult information policy to be realized by cryptographic protocols.

The traditional goals of communication security under intentional errors and disruption are often summarized by the three intuitive notions of confidentiality, integrity, and availability. The overarching term *The Science of Information Integrity* for denoting the general goal of secure communication has been used too.

6.3   Some Concepts

6.3.1   Primitives and Protocols

The goals of cryptographic protocols are not straightforward to describe in a precise way. Researchers in the field of cryptographic protocols find it often useful to distinguish between

Cryptographic Primitives The mathematical operations and functions used in the local cryptographic transformations computed by the communicating parties, for example, one-way hash function, public key trapdoor function, and secret key stream and block ciphers.

Cryptographic Protocols The communication procedures that use cryptographic primitives in achieving some goals, for example, key distribution, entity authentication, confidential information sharing. A more thorough description of the notion of a cryptographic protocol follow in the next subsections.

This distinction of cryptographic primitives and protocols is useful in many situations, such as security analysis. Section 6.5.2 gives some good reasons for this separation of concerns.

We will later show by examples that cryptographic primitives are *necessary but not sufficient* for achieving secure communication. A nice allegorical way of illustrating this fact is the problem of locking a bicycle against theft. You have at your disposal an unpickable padlock , an unbreakable chain to lock your bike, and some fixed rack-in-ground; all perfect "primitives." Then how can you secure your "bicycle system"? It is very likely that you have seen, once in a while, a single front-wheel thoroughly secured to some rack, whereas the rest of the bike has disappeared. Or the frame is locked to the rack while the wheels are no where to be seen. Take a look at Figure 6.1. The padlock and chain components are necessary but not sufficient for security.

FIGURE 6.1
Strong security primitives (e.g. bikelock and rack-in-ground) are necessary but not sufficient for securing a (bike-) system. Evidently, it does matter how things are put together.

6.3.2   Definitions

The basic goal of communication between two parties is to get the message across. This sounds straightforward and simple. However, when starting up the communication, you will almost immediately realize that you will need some extra signaling rules that must be agreed upon and used. When do you start? Is the recipient ready? When is the sender finished? Has the message been received? Where does the sender resume from if a message does not arrive? And so on.

A *communication protocol* is a set of rules that controls the interaction of n communicating entities or parties. Normally, the communicating parties are concurrent processes running in a distributed system of computing devices.

The word *protocol* as a term in this particular communication sense was first used in a memorandum written at NPL, England, in April 1967 titled "A protocol for use in the NPL data communications network."

It is a fun tradition in the field of cryptographic protocols to use personal names like Alice, Bob, and Charlie. This is a most useful and engaging trick when explaining a cryptographic protocol in an intuitive way to people.

However, this anthropomorphism can be very misleading to a nonconsecrated audience or reader because dramatic personal names are put on deterministic soulless computing devices. Normally, there are no person and common sense involved in the protocol execution, just computing machines tirelessly acting according to their programmed instructions.

A more careful distinction between entities, roles, and instances must be made when analyzing cryptographic protocol security. A communicating entity is often named a participant, a party, a principal, a player, or something similar. There are at least two distinct roles to play in a protocol. It might be a simple sender and a receiver, it might be the role of an initiator ("the first mover") and the responder, or the role of prover and verifier, and so on. Similar to the notion of an algorithm, we want to distinguish between the protocol algorithm itself, and an instance of an execution of the protocol. A *run* of a protocol is one instance of a protocol execution where the variables are assigned specific values. The protocol variables will take on specific values for each run of the protocol. A cryptographic protocol may be repeated in several *rounds*. Finally, the logical or physical device of a design or an implementation can usually take on many roles of many different protocols, and perform many runs simultaneously.

The two-party case, $n = 2$, is perhaps the most common model. A *multiparty protocol* is where $n \geq 3$. A possible situation here will be a single sender and multiple recipients, whereas in a more general situation, one has to consider the possibility of both multiple senders and multiple recipients. If we consider *the storage channel*, then we are able to define a special one-party case, where a party or principal can take on both the role of sender (writer) and subsequently the role of recipient (reader).

A *cryptographic protocol* is a communication protocol that includes one or more cryptographic primitives.

A *communication channel*, or just channel for short, is the communication medium enabling the message exchange between the communicating parties. The channel properties can be one-to-one or multicast or broadcast. Normally, we assume a perfect rather than a noisy channel. Some models even assume cryptographic properties of the communication channel, like secrecy channel, or authenticity channel.

A network of communicating parties can be modeled in a natural way as a graph, where the nodes are the communicating parties, and the edges/connections are two-party communication channels.

6.3.3 The Protocol as a Language

We consider the notion of language essential to communication. A language consists of an alphabet of symbols, the syntax of acceptable words of the language, and the grammar of acceptable sentences. The notion of a protocol can be considered analogous to a language. A valid communication message is an acceptable word in the language, and a sequence of message exchanges according to the protocol generates an acceptable sentence in the language. The effect/service to the communicating parties is analogous to the meaning/semantics of the language.

Analogous to a common language, there must be an a priori agreement between the communicating parties about the rules and procedures of the protocol. The five elements of a protocol specification are [2]:

1. The *service* provided to the communicating parties by the protocol.

2. The *assumptions* about the protocol environment, in particular the properties of the channel.

3. The *vocabulary* of messages that can be used in the exchange.

4. The *encoding* of each message in the vocabulary.

5. The *behavior and processing rules* required of the communicating parties by the protocol.

Let us take a closer look at the element aspect of how to describe the processing rules of a protocols. Basically, there is the input–output description and there is the reactive description. The standard method of describing an input–output relation is by a function description $y = f(x)$, and a construction of an algorithm $A_f()$ that realizes the function by transforming the input value to some output value consistent with the function definition, and then stops.

$$y \leftarrow A_f(x), \text{ where } x, y \in \{0, 1\}^*$$

Alternatively, we can make a description based on the behavior of the communicating parties using a reactive finite state machine description $(Q, q_0, M, T(q, m))$, where Q is the set of states, q_0 is the initial state, M is the set of communication messages, and $T()$ is a state transition function with input the current state q and the incoming message m, and the output is the next state q'.

Let us look at an example and introduce some notation based on a message sequence diagram description.

The message sequence diagram of Figure 6.2 shows the normal sequence of messages in the exchange. There are two parties or principals involved in the protocol, Alice and Bob. The diagram has three columns, the left column is for the local computations of Alice, the middle column is for the values transmitted by the channel between Alice and Bob, and the right column shows the local computations of Bob. The sequence of events progresses from the top toward bottom.

Both parties are required to generate random values, where a uniform sampling from a set $\{1, 2, 3, \ldots, p - 1\}$ is denoted by $x \in_R \mathbb{Z}_{p-1}^+$. A variable, say, a marked as $\tilde{a}$ means that the value of a is not locally generated, it comes from the "outside" communications[1]. The variable or list of variables above the communication arrow shows the message content.

---

[1]The value can be "crooked".

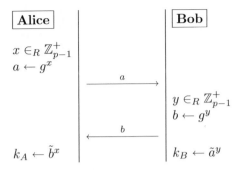

FIGURE 6.2
The Diffie-Hellman key exchange protocol.

A *normal* protocol run can be considered as a composition or chain of input and output values. Private input values such as keys and randomized values can be considered just another input to the function. For the protocol of

The protocol view of principal Alice in Figure 6.2 can be described as a function Π_A

$$k_A \leftarrow \Pi_A(p, g, x, \tilde{b}),$$

and similar for principal Bob

$$k_B \leftarrow \Pi_B(p, g, y, \tilde{a}).$$

And the global view of the protocol can be described as a combination

$$(k_A, k_B) \leftarrow \Pi(p, g, x, y).$$

Figure 6.3 shows another example, now with four messages in the total protocol exchange.

The diagram in Figure 6.3 shows one round of the Fiat-Shamir protocol. The goal of this protocol is that Peggy convinces Victor that she knows the private value s corresponding to a public value p, under a one-way function. The one-way function (the cryptographic primitive) used here is squaring modulo n, where n is the product of two large prime numbers. Computing the inverse, that is, extracting the square roots modulo n is equivalent to factoring n, and factoring is believed to be a computational hard problem. If the public value p can be interpreted as an identifier of some principal, then the protocol can be considered an identification protocol.

Note that there is no intermediate branching in the sequence of events except for error handling. If an error occurs, or a test fails, then the process is assumed to stop after sending an error message. Therefore, a normal run of a cryptographic protocol, as shown here, becomes a completely ordered sequence of local computations.

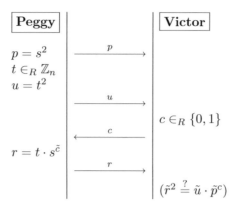

FIGURE 6.3
One round of the Fiat-Shamir identification protocol, where Peggy can show
Victor that she knows the private value s corresponding to the public value p.

6.3.4 Provability

Basically, there are two main aspects or properties of a cryptographic protocol
design that we should seek to verify. The first is *the correctness* for all possible
input values when all parties behave according to the protocol rules. This is
normally not so difficult. Further to this point, we want to show that the
protocol execution can be carried out by efficient algorithms and is practical.

The second aspect is *the soundness* of the protocol, that is, the security
of the protocol principals relative to an adversary. Methods for proving this
have been much more difficult to establish. First, the computational power of
the principals, and the assumptions about their secret and public input must
be asserted. Moreover, we need to model the capabilities of the adversary (see
the next subsection for this). Then we want to state the security claims about
what will happen when attacks are made by the adversary and show that the
claims hold.

In general, we cannot expect to be able to list all possible attacks on the
protocol, so a different strategy is often used: by a "reduction construction,"
starting out from a supposed, but unspecified attack that breaks the protocol
security goal or claim. Then a "reduction" is constructed, essentially showing
that the supposed break implies that an underlying computational hard problem can be efficiently computed. Then the prover denies this result, based on
firm belief, attitude, and accumulated experience he simply refuses that the
computational hard problem at use can be broken. This amounts to using the
logical modus tollens deduction rule of denying the consequent of the reduction implication, where it follows that the antecedent must be false. In our
context, the resulting deduction is that the protocol attack supposition that

we started out with cannot be true. This mathematical method has both its proponents and opponents [3, 4].

6.3.5   Modeling the Adversary

The attacking game can be modeled in several ways. We may assume a weak adversary with limited power, or a stronger adversary with extended power over the communication and the participants. Let us name the adversary Malice.

The weakest adversary model is the passive eavesdropping model of Shannon. Here the adversary Malice has access and reads all communication between the principals, but she cannot take active part in the communication.

If Malice can take an active part in the communication, but still complies with the protocol rules, then the attacker is often characterized as "curious, but honest."

The next step up is to accept that the adversary can combine passive and active attacks. Malice can read all messages between the principals, and she can delete, modify, replay, and generate new messages. Malice "is the network."

A further strengthening of the adversary is to allow Malice to be a legitimate principal as well. She is enabled to perform an "insider attack." This implies that she will have a legitimate identity in the system, possess a set of legitimate cryptographic keys, thereby she is able to initiate interactions with all other principals within the protocol system and use them as "black boxes" in her protocol attack.

In an even stronger attack model, Malice can also be acknowledged the power to take over ("corrupt") other principals, making the opportunity for collusion attacks by all "maliced" principals on the remaining honest principals.

6.3.6   The Problem of Protocol Composition

In general, the question of security of arbitrary cryptographic protocol *composition* is a grand challenge in this research field. We cannot hope to control which cryptographic protocols are running in an open communication network (the Internet), so we must take care that our protocols remain secure within such an environment. One concern is possible interdependencies between security requirements and the execution environment. For instance, messages generated by a run of protocol Π_1 might become useful for an attack on protocol Π_2.

Remember that the communication procedures of a network is conceptually designed as a hierarchy of communication protocol layers, starting at the communication medium with the physical and link layers, the network routing layer above them, then the end-to-end transport, and the application layers on top. All these layers and their sublayers may contain cryptographic protocols.

We can consider several types of protocol composition.

- *Sequential* composition of runs of one or more protocols.

- *Parallel* composition of runs of one or more protocols.

- *Concurrent* protocol composition, that is, allowing arbitrary message interleaving of simultaneous runs of one or more protocols.

Canetti [5] proposed a Universal Composability Framework that addresses this concern. Protocols proven secure within this framework remain secure with other runs of the same or other protocols. However, it is found that there exist functionalities for which it is not possible to construct protocol solutions that adhere to the requirements of this framework.

Since larger cryptographic protocols can be considered a collection of many protocol components, compositional analysis is attractive from an engineering point of view as well. We [6] have developed a symbolic framework for compositional analysis of a large class of security protocols. The framework is intended to facilitate automatic as well as manual verification of large structured security protocols. The approach is to verify properties of component protocols in a multiprotocol environment, then deduce properties about the composed protocol. To reduce the complexity of multiprotocol verification, we introduce a notion of protocol independence and prove a number of theorems that enable analysis of independent component protocols in isolation.

6.4   Protocol Failures

6.4.1   Reasons for Failure

A cryptographic protocol fails if one or more of the claimed security properties of the protocol do not hold against some constructed attack that is within the adverserial model. The seed of protocol failure can be sown in the design, in the development or in the operational process. It can come from:

1. Incorrect design of the cryptographic primitives

2. Incorrect design of the cryptographic protocols

3. Incorrect implementations

4. Incorrect environment assumptions

5. Incorrect operational management

All these concerns must be observed to ensure the correctness and soundness of an operational cryptographic protocol as part of a larger system. The next section will show an example of how secure cryptoprimitives are necessary but not sufficient for designing secure cryptographic protocols.

6.4.2 An Example of Protocol Failure

Figure 6.4 shows a once published and subsequently broken proposal for an end-to-end key distribution scheme, assisted by a key distribution center (KDC) in the network.

A		KDC		B
$r_A \in_R \mathbb{Z}_n$				
$a \leftarrow r_A^3$	$\xrightarrow{\quad a \quad}$	a	$\xrightarrow{\text{request}}$	$r_B \in_R \mathbb{Z}_n$
			$\xleftarrow{\quad b \quad}$	$b \leftarrow r_B^3$
		$r_{AB} \leftarrow a^{\frac{1}{3}} \oplus b^{\frac{1}{3}}$		
	$\xleftarrow{\quad r_{AB} \quad}$			
$\tilde{r}_B \leftarrow \tilde{r}_{AB} \oplus r_A$				
$K_{AB} \leftarrow \tilde{r}_B$				$K_{BA} \leftarrow r_B$

FIGURE 6.4
End-to-end key distribution protocol using a trusted third party in the network and two provably secure cryptographic primitives.

This is a key distribution protocol that is using two provably secure primitives, the "one-time pad" operation $\oplus$, having information-theoretic security, and the RSA cipher using a public exponent $e = 3$, where cube-root extraction is equivalent to modulus factoring.

By picking up all the communicated messages, the attacker's input is (a, b, r_{AB}). The question becomes how to solve the equations $r_B = a^{1/3} \oplus r_{AB}$ or $r_B = b^{1/3}$. Is this cryptoprotocol secure?

- Yes, the cryptographic protocol security appears all right when restricted to an active outsider adversary.

- But if we grant Malice to be an insider that can collude with other principals, then Malice can take over the roles of several principals and act on the inside. The cryptographic protocol example above is not secure if collusion/cooperation of principals can happen. Malice will act as a collusion of two principals in the attack.

How can the attack be carried out?

Figure 6.5 shows one way of attacking the protocol. Observe that the computation of *KDC* can be considered as a "decryption service" for the principals. We assume that the value b has been eavesdropped by C from a previous run, and that C and D are colluding against *KDC*. In particular D will provide C with the random value r_D. Then it turns out that C receives sufficient information to solve the equations and derive the secret key of the run between A and B. The trick by C is to prepare $c \leftarrow r^3 \cdot b$, then C is able to compute the value of the secret key r_B of the eavesdropped protocol run.

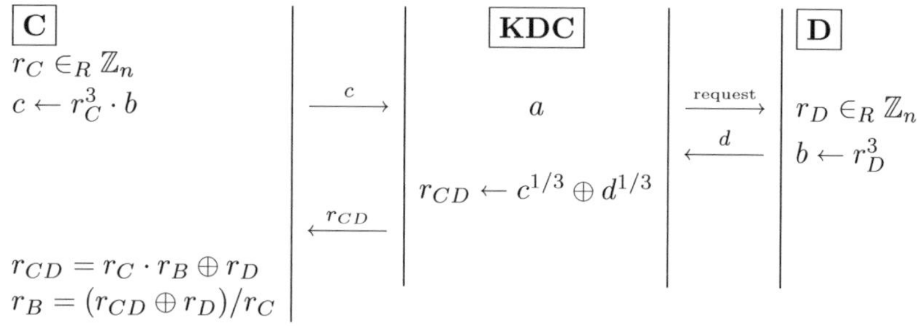

$$\boxed{\text{C}}$$

$r_C \in_R \mathbb{Z}_n$

$c \leftarrow r_C^3 \cdot b$

$$\xrightarrow{\quad c \quad}$$

$$\boxed{\text{KDC}}$$

a

$$\xrightarrow{\quad \text{request} \quad}$$

$$\boxed{\text{D}}$$

$r_D \in_R \mathbb{Z}_n$

$$\xleftarrow{\quad d \quad}$$

$b \leftarrow r_D^3$

$r_{CD} \leftarrow c^{1/3} \oplus d^{1/3}$

$$\xleftarrow{\quad r_{CD} \quad}$$

$r_{CD} = r_C \cdot r_B \oplus r_D$

$r_B = (r_{CD} \oplus r_D)/r_C$

FIGURE 6.5
An active attack on the key distribution protocol.

A simpler alternative will be that C just sends $c \leftarrow b$, but then the *KDC* might detect this as a replay from a previous run of the protocol. The basic flaw detected is in the silent assumption that is made about the input to the *KDC*. The cryptographic protocol designer silently assumed it to be random without structure, but that is only true if the principals follow the protocol.

The wisdom gained is that we must always check to see what happens if random values are chosen nonrandomly and with structure. Check the resilience against sharing private input.

6.5 Heuristics

Let us now review some good advice and recommended best design principles that have been proposed for use in the construction and verification of cryptographic protocols.

6.5.1 Simmons' Principles

Simmons [7] proposes to use the following three heuristic principles when designing cryptographic protocols and checking that they are sound and secure.

Principle 1

Carefully *enumerate all of the properties* of all of the quantities involved, both those explicitly stated in the protocol specification and those implicitly assumed in the setting. Take nothing for granted.

Principle 2

1. *Go through the list* of properties assuming that none of them are as they are claimed or tacitly assumed to be unless a proof technique exists to either enforce or verify their nature.

2. For each possible violation of a property, critically examine the protocol *to see if this makes any difference in the outcome* of the execution of the protocol.

3. *Consider combinations* of parameters as well as single parameters.

Principle 3

- If *the outcome of the protocol can be influenced* as a result of a violation of one or more of the assumed properties, it is essential to then *determine whether this can be exploited* to advance some meaningful deception.

- Protocol failures occur whenever the function of the protocol can be subverted as a consequence of the violations.

The very first cryptographic protocol example we started with, depicted in Figure 6.2, is the Diffie-Hellman key exchange. This protocol is an example for the concern of principle 3, where the outcome can be influenced, but with no meaningful deception.

6.5.2 Separation of Concerns

Divide and conquer is a useful principle in algorithmic design, but also in the more general fashion of science. Descartes recommends

> to think in an orderly fashion, beginning with the things that were simplest and easiest to understand, and gradually and by degrees reaching toward more complex knowledge.

We interpret his guideline as we should try to partition our grand cryptographic protocol problem into subproblems and then to focus on the easier subproblems first.

A direct and natural problem partitioning attempt for cryptographic protocols is to separate the concerns of the crypto-primitives and the cryptographic protocols. What can we solve if we can assume crypto-primitives with perfect security properties? An encryption of the message m under the secret key k becomes abstracted and ideal with the expression $[m]_k$, though a detailed definition of what this notation means in terms of security must be made too.

A simple approach to cryptographic protocol specification between an initiator I and a responder R, based on an idealization of the crypto-primitives, can then go like this:

1. $I \rightarrow R : [N_I, I]_{pk(R)}$
2. $R \rightarrow I : [N_I, N_R]_{pk(I)}$
3. $I \rightarrow R : [N_R]_{pk(R)}$

This describes the Needham-Schroeder public key based mutual authentication protocol. This kind of notation conveys the basic ideas of the protocol in a compact way; nevertheless, this is a very incomplete specification with many inbuilt assumption that can be very deceptive at analysis time. For instance, there are no declarations of types, constants, variables, nor functions. The domains, generation, and computations of values are not shown. And no pre- and post-conditions, the actual goals of the interaction, are claimed.

Figure 6.6 shows a somewhat enhanced description of the same protocol.

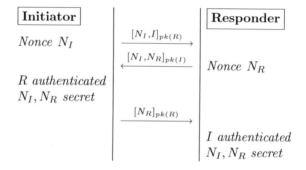

FIGURE 6.6
The Needham-Schroeder public key based authentication protocol.

The main question remains: Is this protocol good with respect to the claimed security properties? No, there exists a man-in-the-middle attack that breaks the claim that the final message must come from the initiator. (Exercise: Try to find it. Hint: Lowe's attack.)

Why is it so difficult to spot the information security bugs in a "three-line program"? One reason is the quite complex underlying security model that is there despite the suggestion of simplicity in the notation and description. Note, for instance, that we have not bounded the number of role instances, nor the number of runs of the protocol. And the assumptions about the communication environment and the outside and inside attackers are neither explicit nor clearly stated.

Figure 6.7 shows a fix for the protocol with respect to the Lowe's attack.

Then, again, is this protocol good now? Note that analytical formalisms do not themselves give design rules directly. So let us review some construction advice in the following section.

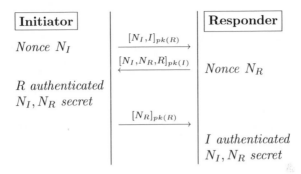

FIGURE 6.7
Lowe's fix of the Needham-Schroeder protocol.

6.5.3 More Prudent Engineering Advice

Abadi and Needham summed up their experience and intuition in a set of heuristic[2] principles for the engineering of cryptographic protocols, based on observations of common patterns in the records of flawed protocol proposals in the 1980–1990s. Their design principles [8] are clearly abstracted from well-known authentication protocol cases, such as the ones presented in the previous subsection; nevertheless, they claim a wider applicability for their principles. The problem focus is along the separation-of-concern approach described in Section 6.5.2, where the cryptographic functions used in the protocols are abstractions with ideal properties.

The two basic principles proposed are

Basic Principle 1: **Explicit Communication**

Every message should say what it means: its interpretation should depend only on its content. It should be possible to write down a straightforward sentence describing the content—or in a suitable formalism available.

For instance, the authentication process of any message must depend only on the data in the message itself or already in the possession of the recipient.

Basic Principle 2: **Appropriate Conditions for Action**

The conditions for a [received] message to be acted upon should be clearly set out so that someone reviewing a design may see whether they are acceptable or not.

The first principle is concerned with the content of the message itself, and the second principle is concerned with the correct reaction to the received message. Curiously, the secrecy and confidentiality of information are not

---

[2]The adjective *heuristic* means that it is aiding or guiding in discovery, inciting to find out.

stated explicitly in the principles proposed, though it is clearly a supposition that the protocol must not leak private parameter values.

Nine more principles are proposed to be derived from the first two ones. Whereas we simply list the principles here, Ref. [8] discuss each principle with examples.

3. If the identity of a principal is essential to the meaning of a message, it is prudent to mention the principal's name explicitly in the message.

4. Be clear about why encryption is being done.

5. When a principal signs material that has already been encrypted, it should not be inferred that the principal knows the content of the message.

6. Be clear about what properties you are assuming about nonces.

7. If a predictable quantity is to be effective, it should be protected so that an intruder cannot simulate a challenge and later replay a response.

8. If timestamps are used as freshness guarantees, then the difference between local clocks at various machines must be much less than the allowable age of a message.

9. A key may have been used recently, for example, to encrypt a nonce, and yet be old and possibly compromised.

10. It should be possible to deduce which protocol, and which run of that protocol, a message belongs to, and to know its number in the protocol.

11. The trust relations in a protocol should be explicit, and there should be good reasons for the necessity of these relations.

Summing up, bear in mind that these 11 principles reviewed here are clearly not sufficient in the broader context of all the information policies and possible protocol goals discussed in Section 6.2 above.

6.6 Tools for Automated Security Analysis

We have now seen that making informal statements and claims about cryptographic protocols is not reliable, so several researchers have been thinking that automated tools based on formal logic may be worth a try. Modeling based on formal logic holds promise of being a useful tool in analyzing the security of cryptographic protocols, where at least the verification of proofs for the security claims can be mechanized.

One problem with this approach is that a different formalism seems to be required to capture each type or class of security claim. Furthermore, the mechanized proofs tend to be long and tedious, even for "obvious" security properties. And we must remember that the tools are only as relevant to the security analysis as are the formal abstractions they are built on, which means that a tool cannot show the absence of flaws in a cryptographic protocol. If a tool reports no flaw, we ought to be more convinced about the protocol than without the tool result, but it is not the final verdict in the full context of Section 6.4.1.

On the positive side is that the analysis of cryptographic protocols by formal logic can lead to a better understanding of the problems at hand and the properties that any solution to it must have. There exist many methods and tools for the analysis of cryptographic protocols. *Scyther* is one interesting formal analysis method tool that is readily available for download and use. It is a specialized model checker for authentication protocols that tries to emulate simple theorem proving methods. It uses backward trace search in a finite state model, so the computation is normally very fast and is guaranteed to stop. There is a fairly easy language for protocol specifications and claims. The Scyther tool illustrates the attack scenarios found by a graphical diagram. You can read more about the various formal method approaches to security protocol analysis and verification in the books referenced in the Further Reading Section 6.7.

6.7 Further Reading and Web Sites

Simmons describes how cryptographic protocols fail even though the cryptographic primitives are excellent [7]. The example of a cryptographic protocol failure in Section 6.4.2 is adapted from [9].

Abadi and Needham [8] proposes principles for best practice for designing cryptographic protocols. The principles are neither necessary nor sufficient for correctness, but helpful in that adherence to them can prevent a number of published errors. Examples given show the actual applicability of these guidelines.

The Scyther software tool can be downloaded from `http://people.inf.ethz.ch/cremersc/scyther/`. A repository of security protocol descriptions is available at `http://www.lsv.ens-cachan.fr/Software/spore/table.html`.

The graduate level textbook *Modern Cryptography: Theory and Practice* [10] contains an extensive and advanced treatment of cryptographic protocols. The book *Protocols for Authentication and Key Establishment* [11] is a very comprehensive research level treatment with lots of references, but it includes tutorial material as well. *The Handbook of Applied Cryptography* [12] covers many issues of cryptographic protocols and is fully available online at http://www.cacr.math.uwaterloo.ca/hac/. Chapter 10 presents identification protocols and their practical constructions.

There is an exciting world of cryptographic protocols that go beyond the classical goals of key distribution and channel security, such as blind signatures, digital cash and credentials, sender anonymity, oblivous transfer of information, commitment protocols, zeroknowledge identification, secret sharing and multiparty computations, and much more. The research literature is the best source for more information on these topics. Start by checking out the web site www.iacr.org.

Bibliography

[1] W. Diffie. The first ten years of public-key cryptography. *Proceedings of the IEEE*, 76(5)(5):560–577, May 1988.

[2] G. J. Holzmann. *Design And Validation Of Computer Protocols*, volume 94. Prentice Hall, 1991.

[3] N. Koblitz. The uneasy relationship between mathematics and cryptography. *Notices of the AMS*, 54(8):972–979, 2007.

[4] I. Damgård. A "proof-reading" of some issues in cryptography. *Automata, Languages and Programming*, LNCS 4596:2–11, 2007.

[5] R. Canetti. Universally composable security: A new paradigm for cryptographic protocols. Technical report, Cryptology ePrint Archive: Report 2000/067, Revised 13 Dec 2005.

[6] S. Andova, C. Cremers, K. Gjøsteen, S. Mauw, S. F. Mjølsnes, and S. Radomirovic. A framework for compositional verification of security protocols. *Information and Computation*, 206(2–4):425–459, 2008.

[7] G. J. Simmons. Cryptanalysis and protocol failures. *Communications of the ACM*, 37(11):65, 1994.

[8] M. Abadi and R. Needham. Prudent engineering practice for cryptographic protocols. *Software Engineering, IEEE Transactions on*, 22(1):6–15, 1996.

[9] Choonsik Park, Kaoru Kurosawa, Tatsuaki Okamoto, and Shigeo Tsujii. On key distribution and authentication in mobile radio networks. In *EUROCRYPT '93: Workshop on the Theory and Application of Cryptographic Techniques on Advances in Cryptology*, pages 461–465, Secaucus, NJ, USA, 1994. Springer-Verlag New York.

[10] W. Mao. *Modern Cryptography: Theory And Practice.* HP Professional Series. Prentice Hall PTR, 2004.

[11] C. Boyd and A. Mathuria. *Protocols for Authentication and Key Establishment.* Springer Verlag, 2003.

[12] Alfred J. Menezes, Paul C. van Oorschot, and Scott A. Vanstone. *Handbook of Applied Cryptography.* CRC Press, 2001.

7

Public Key Distribution

S. F. Mjølsnes

Department of Telematics, NTNU

CONTENTS

7.1   The Public Key Distribution Problem

The problem of distributing and managing cryptographic keys is simpler with public keys than with secret keys. Obviously, a secret key must be kept confidential, but not so for a public key. We do not need to worry about the confidentiality of a public key. Rather it should become as publicly and readily available as possible. Nevertheless, public keys do not escape the other requirements of cryptographic key distribution. These are the requirements of authenticity of origin, the integrity of the key value, and the validity of usage.

In their foundation paper on public key systems, Diffie and Hellman imagined a central server that maintained the public keys [1].

> [The public key distribution system's] use can be tied to a public file of user information which serves to authenticate user A to user B and vice versa. By making the public file essentially a read only memory, one personal appearance allows a user to authenticate his identity many times to many users.

Now having the advantage of hindsight, we may say that they underestimated the technical challenges of a practical set up of a public key directory.

Let us consider the printed telephone book that is now becoming replaced by online database management systems. The content of the telephone book presents a directory of name and telephone number for some specific region or municipality. The list of records is sorted by surname in lexicographical order. You are expected to know the municipality of the person you are trying to call. Even though you know the proper spelling of the last name of the person you want to call, it is not unusual that there are two or more people with the same name. So which telephone number do you select? You might try to connect to one number at a time, and then hope for some validation in the beginning of the conversation that you have reached the correct callee.

So can a similar electronic registry work for distributing public keys? There are several issues and remarks to be made:

1. There are always printing errors in a telephone book, whereas errors in the public key simply must not be.

2. The telephone book is issued by a telephone company that provides the service, and the correct telephone numbers are an essential part of the provisioning of that service. The public key is not thought to be specific to one communication service. Moreover, the private and public keys are generated by the users in the classic public key model, not the service providers. This difference has turned out to be a significant practical stumbling block.

3. The caller must know the location of the callee in order to look up the correct telephone number. Now mobile telecommunication has agreed on an IMSI (International Mobile Subscriber Identity) for the interoperation between mobile operators.

4. If the wrong number is called, then normally it can be cleared up in the beginning of the conversation because of personal voice communication validation of situational information. This does not make sense and is not allowable in the security model of public keys.

5. Subscriptions are entered and left on a continuous basis, whereas normally a telephone book release is on an annual basis. Hence, the update problem of a dynamic database. The question of validity and revocation of a public key is a major issue in Public Key Infrastructures.

6. The telephone number is a network address for a specific telecommunication service, so it is service specific. Certainly, a public key might be associated with a specific service, but that was not the early vision of public key usage.

Therefore, the similarity between a list of telephone numbers and list of public crypto-keys does not hold against closer examination at all. The comparison fails in most respects actually.

There has been substantial critique against the idea of a universal architecture of Public Key Infrastructure (PKI) over the years. Moreover, many ambitious national and international projects for realizing a common public key infrastructure have been carried out with less than success. Many issues for debate exist; some contentious propositions that can be made are

- The original notion of Public Key is fully distributed with a minimal of centralized online resources, whereas a practical process of public key assignments must have an online revocation registry to control validity periods. Therefore, the difference between a normal online authorization system using symmetric encryption and public keys is nil in this respect.

- The distributed, but still centralized hierarchical architecture of PKI does not scale well in global communications across the decentralized Internet; therefore, the interoperability problems showing in practice are not surprising.

- The competitive nature of business organizations is typically found to be incompatible with the innate liability and legal requirements typically brought forth by public key certifications.

- National digital signature laws command the equivalence between a handwritten signature by pen and a digital signature by a cryptographic key, whereas in strict technical terms, no such similarity can be observed. The impetus of these laws stems from wishful storytelling rather than computer engineering.

- The human behavioral needs for flexibility, tentativity, and ambiguity in the interpretation of signatures and agreements are not taken into account.

7.2   Authenticity and Validity of Public Keys

This scenario should illustrate the problem of public key distribution. Alice has written a technical paper that has been submitted and accepted by a publisher, here represented by the editor Bob. They communicate by e-mail over internet. Bob requests a copyright transfer agreement be signed by Alice before he can publish the paper. So Alice inputs the agreement text and computes a digital signature with her private key and her preferred algorithms, then appends the resulting value to the agreement text. She also includes her public key in her e-mail response to Bob.

In principle, this enables Bob to verify the provided digital signature of the agreement. However, why should Bob be convinced that the provided public key and the resulting signature belongs to Alice? Remember that although the public key presumably has been generated by Alice, the public key itself is just a nonsensical value or text-string with no link to Alice the person or her name and identity. A handwritten signature contains both the name of the signer, and a personal graphical imprint of the name. The paper of a handwritten letter binds the content and the signature into an integrated whole. And the letter contains similarity of the graphical characteristics of the ink and the handwriting style. None of this is present in the digital public key. The only remaining link to Alice is in the actual e-mail text that Bob received, but then this could have been equally convincing without the digital signature.

Imagine then that Bob met with Alice at her office at some earlier occasion, where she handed her public key to him, say, by a piece of paper. So the publisher has registered Alice's public key "on file" for reference already, and Alice's claim that this is her public key is corroborated by this prior registration. Unfortunately, this mode of distribution brings us back to the practice of key couriers in classic cryptographic key distribution.

Then there is always the general problem of maintaining the consistency between what is registered at a certain point in time and the current actual situation. Some time after Alice handed Bob her public key, she became uncomfortable about an incident that might have resulted in revealing her private key. She decides to be on the safe side and generates a new private-public key pair for her future use. However, now Bob's registration of Alice's current public key becomes invalid. And what will come of the agreements and contracts Alice has signed with her compromised and discarded signature key?

7.3 The Notion of Public Key Certificates

7.3.1 Certificates

A birth certificate is a document issued by the authorities that attest the birth date and place of a named individual. The certificate can contain the child's name, the date and place of birth, gender, the name and address of the parents, the time of registration, and the registration authority. The birth certificate can later be used to verify the age and name of an individual.

A passport is another example of a certificate issued by the national authorities for the purpose of international travel across borders. It certifies the name and nationality of its holder, date and place of birth, gender, and personal characteristics such as height, the color of eyes and hair, and includes a full face photo. The passport is officially issued with a validity period.

Our daily business and activities are full of certification documents, such as driver's license, credit card, and identity card. In general, we can say that a certificate contains a statement or claim about one or more objects and their attributes, together with a third-party endorsement that the claim is valid. In particular, we can certify that a public key is associated with some named person or organization, a public key certificate.

7.3.2 Public Key Certificates

We will analyze the concept of public key certificates now. Let us start out with the simplest data structure one can imagine for a public key certificate, where a name is bound to a public key value with a digital signature.

```
((Name, Public_key), Certifier_Signature)
```

How can the signature of this certificate be verified? Well, we need the public key of the certifier.

It may be that it is obvious who the certifier is, by a given context. In particular, if the certifier takes on the verifying role, then it is reasonable to assume that the verification key is readily available. The same principal takes on both the certifying and verifying role. This setting may have certain advantages if off-line verification is necessary, but simply making an authenticated connection and a lookup at a centralized directory of

```
{(Name, Public_key)}
```

is a very attractive alternative if an online communication service is available already.

It may not be obvious who the certifier is and the corresponding public key. One solution would be to apply the same public key certificate method again, but now for the certification of the certifier's public key. The data structure becomes

```
((Certifier_Name, Public_key), OtherCertifier_Signature)
```

which is essentially the same situation as what we started with, so we are going in circle here.

The verification process corresponds to traversing a linked list of certificate nodes.

```
((Name, Public_key, Certifier_Name), Signature)
```

where the Certifier_Name is the pointer to the next node in the list. See Figure 7.1 for an illustration of this.

There are three possibilities for a finite list of certifications. One possibility is that the last node contains a null pointer, meaning that the ultimate certifier is not certified. A variant of this is that the ultimate certifier points to itself, a Napoleonic crowning act of this public key by a self-certifying signature! A third possibility is a circular list where the last node points to the another node, for instance, the first node in the list.

Some thinking should convince you that all three options are just shifting, not solving, the problem of certification because the final certification is not anchored to anything but itself. A crucial assumption is needed for this linked list of certification to make sense, and to escape the fallacy of circular reasoning. One approach is to ensure that the validity of at least one of the name and public key bindings in the linked list can be established by other means.

A disclaimer

Information security is a tricky subject and it is easy to overlook implicit assumptions. Here we note that a public key certificate gives no guarantee that:

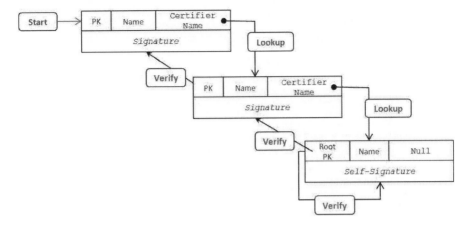

FIGURE 7.1
A signature chain of three public key certificates, including the root certificate with a self-signature.

- the private key is kept secret, and that no other than the "associated owner" has been able to use it.

- the private key is kept secret, and that no other than the "associated owner" can or will be able to use it.

7.3.3 Certificate Data Structures

The process of constructing and standardization of a comprehensive and practical data structure for a public key certificate in an internet environment has been a long and winding development. The main fields of X.509 v.3 data structure standard are the following:

Version Identifies the version of this data structure.

Serial Number A certificate instance identifier assigned by the issuer.

Algorithm ID Identifier of the signature algorithm used to create the signature.

Issuer Name A unique name of the certificate issuer.

Validity Period Consists of two dates, the date when the validity period begins, and the date when the validity period ends.

Subject The name associated with the public key.

Subject Public Key Info This field contains two subfields.

Public Key Algorithm Same as Algorithm ID.

Subject Public Key The public key associated with the Subject.

Extensions The first two versions of the X.509 certificate data structure fell pitifully short of satisfactory for most applications, so the pragmatic solution was to introduce an optional extension field that can contain both standardized and issuer-specific subfields defined in version 3. Some of these are

Authority Key Identifier An identifier of the issuer's public key that must be used to verify the certificate.

Subject Key Identifier An identifier of the public key contained in the certificate.

Key Usage It can be used to identify (or restrict) the services that the certified public key can be used for.

CRL Distribution Point The address of where the Certificate Revocation List associated with this certificate is located.

Certificate Signature Algorithm The same as Algorithm ID field. An Object Identifier (OID) is a unique representation allocated to an signature algorithm (for instance, SHA-1 with RSA is 1.2.840.113549.1.1.5).

Certificate Signature The digital signature value of the certificate.

Figure 7.2 shows an example of a certificate coded according to X.509 public key certificate issued by someone.

7.3.4   Chain of Certificates

The generation and distribution of digital certificates made by some certification authority that shall vouch for the link between public keys and principal names do not solve the security problem; it only shifts the authenticity problem to another key: the certification key. The public key of the certificate issuer is needed to verify the certificate. Although this verification key can be included along with the certificate data, the verifier needs to establish that this is the correct and valid public key of the certificate issuer. In principle, the certificate could be generated by anyone. Therefore, we arrive at basically the same problem as we started out with, but now the object of doubt is the certificate issuer's verification key. More generally, we can introduce additional certification of the certifiers, then certify the certification of the certifiers, and so on. Clearly, a *chain of certification* has to end by some other means than a digital certificate if we are to solve the problem of public key validity. This public verification key is often called *the root key*. Take a look at Figure 7.1 again.

One way to end the certification chain is to establish a *self-signed certificate*. This means that at the end link of the certificate chain the certificate

Certificate: Data: Version: 3 (0x2) Serial Number: 204 (0xcc) Signature Algorithm: sha1WithRSAEncryption Issuer: C=NO, ST=Sor-Trondelag, L=Trondheim, O=NTNU, OU=Institutt for Telematikk, CN=SIE5040 - Personal Test Certificates Validity Not Before: Feb 6 12:03:49 2003 GMT Not After : Oct 14 12:03:49 2003 GMT Subject: C=NO, ST=Sor-Trondelag, L=Trondheim, O=NTNU, OU=Institutt for Telematikk, CN=Stig Frode Mjolsnes/Email=sfm@item.ntnu.no Subject Public Key Info: Public Key Algorithm: rsaEncryption RSA Public Key: (1024 bit) Modulus (1024 bit): 00:a8:a6:a6:be:ab:80:96:55:8c:e0:42:c1:9f:30:
12:fc:56:03:30:14:46:6f:fe:ca:3f:5e:fd:c9:11: d6:9e:88:f8:96:61:66:a0:bb:85:91:37:41:7e:2d:
e6:08:05:0f:0b:1e:51:f3:61:07:57:a0:8b:e1:93: 2a:f4:a4:ea:b4:66:cd:0b:1e:f9:ab:0b:e2:74:4b:
31:ae:66:f1:22:7d:94:04:99:54:1a:22:de:be:96: 45:ff:40:22:91:99:42:4c:d2:d2:bf:6e:13:df:57:
6d:12:bf:4c:d3:fa:fd:e5:1b:2d:d5:35:b7:4d:48: 2e:b2:d1:95:48:c1:eb:a9:0d Exponent: 65537 (0x10001) X509v3 extensions: X509v3 Subject Key Identifier: 06:93:D7:7B:86:E8:8A:B4:14:78:19:69:1C:C4:80:D5:73:43:92:98 X509v3 Authority Key Identifier: keyid:78:55:91:33:47:9B:37:DD:49:C5:E2:DC:E6:35:6E:74:8B:1E:60:10 DirName:/C=NO/ST=Sor-Trondelag/L=Trondheim/O=NTNU/OU=Institutt for Telematikk/CN=SIE5040 - Root CA Certificate serial:01 X509v3 Key Usage: Digital Signature, Non Repudiation, Key Encipherment X509v3 Extended Key Usage: E-mail Protection, TLS Web Client Authentication X509v3 Issuer Alternative Name: URI:https://sie5040.item.ntnu.no/cgi-bin/pyca/get-cert.py/PersonalCerts/ca.crt X509v3 CRL Distribution Points: URI:http://sie5040.item.ntnu.no/cgi-bin/pyca/get-cert.py/PersonalCerts/crl.crl X509v3 Subject Alternative Name: email:sfm@item.ntnu.no Netscape Comment: SIE5040 - Personal Certificates Netscape Base Url: https://sie5040.item.ntnu.no/cgi-bin/ Netscape CA Revocation Url: pyca/get-cert.py/PersonalCerts/crl.crl Netscape Revocation Url: pyca/ns-check-rev.py/PersonalCerts? Netscape Renewal Url: pyca/ns-renewal.py/PersonalCerts? Netscape CA Policy Url: ca/policy/PersonalCerts-policy.html Netscape Cert Type: SSL Client, S/MIME Signature Algorithm: sha1WithRSAEncryption 88:55:f9:9a:01:d5:20:9c:b6:f5:b5:f4:99:f4:ac:fc:50:dd:
41:98:50:54:d3:e6:5a:73:25:31:1e:4d:99:bc:4b:55:45:c7: 4e:e1:94:a0:d3:4d:83:48:9a:a3:ca:a1:cb:67:b7:57:9e:9b:
ab:a6:e6:3c:d9:93:dc:4d:fa:62:6f:01:66:5f:60:c4:69:56: 87:76:6b:09:34:a5:ba:da:ed:de:95:07:c3:dd:54:e1:1d:e3:
b2:e4:b4:d6:ab:9d:d7:1e:6e:5d:d0:08:21:c6:b5:46:2a:31: 28:91:0e:fd:9d:81:c7:7f:be:8a:61:2d:7f:51:9e:eb:41:a4:
eb:e3

FIGURE 7.2

An example of a X.509v3 certificate of an RSA public key of length 1024 bits.

verification key will equal the certified public key itself. A self-referential statement like: "I can be trusted because I am trusted." The issuer of a self-signed certificate states that there is no reason to go beyond the issuer itself to attest to the correctness of the public key. Take a look at Figure 7.1 again for the illustration of this.

Another way to end the certification chain is by making the public key available by many independent channels; then the consistency of the the public key information over the channels in time corroborates the validity and correctness.

A third way is to establish a root public key is by embedding it into a hardware device by the manufacturer, and assume that the manufacturer is providing the correct root key for the verification of the chain of certificates.

Once the chain of certificates has been established by the pointers in the certificate data structure, then the verification process can start with the root key. The first public key certificate in the certificate chain is verified by the root key. Then the next certificate in the chain is verified by the first public key, and so, until the last certificate contains the public key that is going to be used in the application. Alternatively, the verification process can start from the public key in question and traverse the list toward the root public key.

7.4 Revocation

7.4.1 The Problem of Revocation

The public key certificate includes a field that holds information about the validity period of the certification. This consists of two dates, the date when the validity period begins, and the date when the validity period ends. The certificate expires when the validity period ends. This sets a time limit for the cryptoperiod—the time period of active use—of the public key.

However, events might occur that require terminating the validity before the certificate validity period ends. The certificate binds a public key to a user name. This name can change for many external reasons. A legal name can change because of marriage, or the user's privileges linked to a specific user name have been revoked, for instance, because of change of job or e-mail address. The status of the public key can change too. The worst case is that the corresponding private key has been compromised, lost, or deleted by error. Or results of cryptanalytic progress necessitates public key system upgrade with immediate consequences.

A physical certificate, for instance a driver's license, is revoked by confiscation of the physical object. This is not possible for digital certificates not bound to physical objects.

There are three possible directions for a solution to this revocation problem. The first approach is to limit the certificate validity to such a short time period that the need for revocation vanishes in practice. Another direction to go is to establish an online Certificate Status directory, where the certificate revocations are listed. It follows that the user must be able to locate the directory, and a lookup must always be done in order to be certain about the validity of the public key the user is about to use.

The third direction is to issue another certificate nullifying already issued public key certificates. This approach is normally denoted Certificate Revocation Lists (CRL). A CRL is a list of revoked certificates signed by some authority within the PKI system, normally the same authority that has issued the public key certificates. The idea is then to periodically publish this list by some distribution service.

7.4.2 The CRL Data Structure

The main fields of the X.509 version 2 CRL structure are

Version Gives the version of the data structure.

Signature Identifier (OID) of the digital signature algorithm used to calculate the digital signature on the CRL.

Issuer The name of the issuer and signer.

This Update The time of this issue.

Next Update The time of next issue.

Revoked Certificates A list of certificate references, structured as

> **Certificate Serial Number** The certificate instance identifier assigned by the issuer.
>
> **Revocation Time** The time in which the certificate has been invalidated.
>
> **Extensions** Optional extensions include Reason Code, Certificate Issuer, and Hold Instruction Code supporting temporary suspension of a certificate.

Authority Key Identifier The identifier of the verification key for the signed list.

Issuer Alternative Name One or more alternative name forms associated with the CRL issuer.

CRL Number The serial identifier assigned by the CRL issuer.

More extensions There are many more possible fields to the data structure.

Certificate Signature Algorithm Identification of the signature algorithm for this CRL certificate.

Certificate Signature The digital signature value of this CRL certificate.

7.5 Public Key Infrastructure

A distributed computer structure and system that creates, distributes, and manages the public key certificates is commonly referred to as a Public Key Infrastructure (PKI). The general vision of PKI, as brought forward by telecom, governments, and international standard organizations, is a system that can bind user names to public keys. There are a variety of motivations for building such systems. The public sector wants, for efficiency and cost reduction, to shift from paper forms and documents to digital transactions with its citizens and companies. Public key based security, such as digital signatures, is seen as a major enabler in this transition to digital means.

The banking and financial sector has made extensive efforts over the years in trying to build some PKI into their business models and operations. Traditionally, the security model of banks has followed the unilateral model. In other words, the customer is presented with the policy that "by using our system you must trust that the bank will take care of the security of the banking system for all parties involved." This kind of policy clashes with the natural notion of a private key held by each customer. It follows directly from the bank's policy that the bank should manage the private key of the customer too. Something that voids the advantage of the bilateral model of public key system in comparison to a symmetric key system. One case in point is the Scandinavian web-only bank turned away from client-side public key certificates in its web bank access control after using this for some years.

Many smart card based access control systems have onboard public key services, mostly limited to the realm of a single organization or company. Mobile network operators can provide public key services based on functionality in their SIM cards.

The main entities of a PKI system are

Registration Authority This RA entity will receive certification requests, verify the authenticity of the request, and initialize the signature process. A centralized RA is not convenient for a large distributed organization where the authentication procedure is based on physical presence. It is natural then that the RA can be a collection of local registration authorities (LRA).

Certification Authority This CA entity will generate the public key certificates based on the requests arriving from the RA, using the certifier's private signature key.

Certificate Repository This CR entity runs the database of public key certificates. This is where the certificate users will obtain the certificates they find necessary in the process of public key authentication.

Certificate Revocation Repository The CRP entity will maintain the revocation list (CRL) of certificates. This entity can be used as an online service to validate a public key. It can also be used as a point of distributing the whole CRL to local servers.

If we got multiple CA systems running, it might be useful to form them into a structured cooperation. This CA organization may form a strict hierarchy, or an easily expandable tree-structured ordering, or it may form a looser federation of certificate servers.

Cross-certification is the process of relating two previously unrelated systems, say, CA1 and CA2, so that users in both systems can cooperate using their existing public keys. CA1 will recognize CA2 by issuing a certificate for CA2 public key, and CA2 will issue a certificate for the public key of CA1.

7.6   Identity-Based Public Key

Normally, a public key is a random string because it is derived by a one-way transformation from the private key. Of course, the private key must be generated at random. So there is no information in the public key itself that we can link to a person or the identity of a device. One purpose, among many, of giving names to people, animals, cars, and things is that we may easily refer to them in our interactions. For many circumstances, a name is already established, introduced to the surroundings and well known by usage. A third-party certification of the name is not necessary in daily business. And some sort of official documentation that link the name of a person to the actual person is likely to exist already. The driver's license, the passport, the birth certificate, and more, are all trusted third-party certification of the link between the name and the physical person.

The basis for *identity-based public key* is that the name of the person is already well established in the context of usage for the public key. The "name" may include first name, middle name, last name, address, telephone number, e-mail address, account number, customer number, social security number, photo, fingerprint, and other personal identifying information. It follows that if Alice changes some piece of identifying information included in her identity-based public key, say, her address, then this will of course effect her public key which must be updated, and her outdated public key must be revoked. On the other hand, this variation of public information creates the opportunity to use context-dependent public keys, say, the social security number will only be used in transactions with the public authorities. Or the public key might be an author's pseudonym. If too little identifying information is used for the public key, then uniqueness becomes an issue. John Smith is a case in point.

The security of public key systems are based on the intractability of computing the private key when given the corresponding public key. This works because the public key generation will start with generating the private key and then computing the public key. Now for identity-based public key systems we start out with the public key; this is what is given. Then it should be computationally intractable to derive the corresponding private key. How can this be solved?

Enter a trusted third party T that is set up to compute the secret transformation from a given public key to the corresponding private key, and then issue this private key securely to the user. T generates its own private key, by which T is enabled to compute the private keys for the users in the system. T must authenticate the claimant of a public key to be the correct person before releasing the private key to the person. This authentication problem is comparable to the Certificate Authority's problem of verifying that a public key belongs to the claimant.

The problem of revocation of a public key remains and seems even more problematic than for a certificate-based distribution of public keys. How can the new public key differ from the old public key if there is no external reason to change the identifying information. The name, address, and social security number are the same, not to speak of the difficulty with biometric data. Some suggestions can be to include a serial number, the date of issue, or some other system-defined data. But these data are not a priori linked to the identity of the person, hence we are on a path heading back to the need for explicit authentication within the system, which is where we originally started.

It is vital for confidentiality to use the correct public key. It does not seem that we can escape the need for a public directory of valid public keys for encryption. A signer will presumably already have the valid private key available for generating a digital signature, and the corresponding public verification key can be included in the signed message, so here it seems that identity-based public key systems do have an advantage.

7.7 Further Reading and Web Sites

The book *Understanding PKI* [2] describes concepts, standards, and deployment considerations in further detail.

The EuroPKI conference [3] series focuses on all research and practice aspects of public key infrastructures, services, and applications, and you can find many topics for further study here. This annual conference series started in 2005, the proceedings book are all published on Springer and can be found online at SpringerLink http://www.springerlink.de. An overview site is still operational at www.europki.org. A series of annual PKI R&D workshops were organized from 2004 to 2006 by Internet2 consortium and the proceedings are available online at http://middleware.internet2.edu/pki03/.

The openSSL.org is the widely popular toolkit of open source toolkit that provides a lot of cryptographic primitives that you need for setting up a PKI of your own. The software is available for linux, mac, and windows. The web page at http://www.eclectica.ca/howto/ssl-cert-howto.php describes how to use the openSSL to establish a root certificate authority. The Open CA at http://www.openca.org/ is a collaborative open source project building PKI software. The CAcert.org is a nonprofit Certificate Authority that issues digital certificates to the public at no cost.

Inspect inside your favorite web browser for an overview of root certificates it holds from trusted international digital certificate issuers.

There are many governmental sites dealing with PKI. Many countries have directed laws for the legal acceptance of electronic and digital signatures, following the European Community Directive 1999/93/EC and U.S. Federal Law of the Electronic Signatures in Global and National Commerce Act. The 30th of June has become National ESIGN day: http://www.youtube.com/watch?v=_ji6oxhbdP0. NIST gives a good overview of the standards and activities on the federal level in http://csrc.nist.gov/groups/ST/crypto_apps_infra/pki/index.html.

The influential digital certificate standard X.509 grew out of the X.500 Directory Services series of recommendations, all due to the ITU, see http://www.itu.int/rec/T-REC-X.509. The work group PKIX of IETF that focuses on PKI standards, in particular related to the X.509 standards. `http://www.ietf.org/dyn/wg/charter/pkix-charter`. OASIS is a nonprofit consortium for open standards, and there is a long and impressive list of international PKI-related standards at `http://www.oasis-pki.org/resources/techstandards/`. Actually, the web pages of OASIS give a wealth of references and readings on PKI topics. `http://www.oasis-pki.org/resources/index.html`

Bibliography

[1] W. Diffie and M. Hellmann. New directions in cryptography. In *IEEE Trans. on Info. Theory*, volume IT-22, pages 644–654, 1976.

[2] C. Adams and S. Lloyd. *Understanding PKI: Concepts, Standards, And Deployment Considerations*. Addison-Wesley Longman Publishing Co., Inc. Boston, MA, USA, 2003 (2.edition).

[3] Stig F. Mjølsnes, editor. *Public Key Infrastructure. 5th European PKI Workshop: Theory and Practice, EuroPKI 2008.*, volume 5057 of *Lecture Notes in Computer Science*, Berlin, 2008. Springer-Verlag.

8

Wireless Network Access

S. F. Mjølsnes and M. Eian

Department of Telematics, NTNU

CONTENTS

8.1   Introduction

Communications carried by electromagnetic radio waves have been used for more than a hundred years now.[1] And here we are continuously improving the technological means and organization for how to do wireless communications more efficiently and faster, both in short and long ranges. The term wireless communication includes the use of radio and microwave frequencies, infrared and higher light frequencies. Acoustic waves are wireless too but are not considered here.

One obvious advantage of wireless is that it is wireless, it permits communication without rolling out and connecting cables, electrical conductors, or other waveguides. Microwave links cross geographical barriers like valleys and mountain ranges. Communication satellites can relay signals from one continent to another.

While a cable has two well-defined termination locations, the radio wave channels do not. The waves can be broadcasted, reflected, and diffracted so that the communication signal can reach a large geographical area and cover

---

[1]In 1887, Heinrich Hertz was the first to observe and demonstrate that electromagnetic waves could propagate in "the aether". A few years later, Guglielmo Marconi successfully carried out further experiments, aiming at "Telegraphy without Wires".

a large number of receivers. Observe that the term *broadcasting* has become synonymous with electronic mass media.

Moreover, mobility becomes possible and is a major feature of wireless communication. The transmitter and receiver terminals can change location and move about while being connected. We all know the success and take part in mobile cellular telephone communications with roaming terminals.

Even the core network of communication nodes does not need to be tethered by cables, but can be wireless and adapt to the location of wireless terminals. Actually, the distinction between network terminals and intermediary nodes is not a necessity anymore, so networks of communication nodes can form ad hoc configurations, where all nodes contribute to the total network connectivity by forwarding data for other nodes in a dynamic fashion. Vehicular ad hoc networks is the term used for the development of technology that provides data communication among vehicles and between vehicles and roadside equipment and to be deployed in the near future.

What are distinguishing factors between wired and wireless communications when we consider information security?

Wiretapping needs physical access to the wire, whereas wireless does not, so the geographic area for potential successful eavesdropping becomes much wider. Radio signals are not confined to physical perimeters and barriers, such as inside of a building wall or a fence. The sender broadcasts the radio waves over a wide field, which is actually an important feature in many services, such as television and radio programming with many receivers geographically distributed. Some will even claim that they have rightful ownership to radio waves that enter their land and property!

Whereas monitoring and detection of physical intrusion to a cable is possible, this is hardly even a theoretical possibility with passive eavesdropping of radio communication. This makes eavesdropping on wireless much easier than wiretapping, not demanding physical access and little risk of detection.

Neither are active channel attackers confined to a single location. Active radio transmitters are possible to locate by radio-direction-finding equipment and cross-bearing methods, but this requires active surveillance, the ability to distinguish between authorized and nonauthorized, and carry out a response in realtime. For instance, the source of intermittent disruptions will be hard to discover.

So using wireless for two-party confidential communications is like two people trying to have a private conversation at some distance at a crowded marketplace by means of megaphones. Everyone is able to listen, and the most interested listeners can even move about to obtain improved listening conditions because it is possible to get a directional sense of the location of the communicators.

We can easily recognize and distinguish human voices, and we often use this authentication skill in our telephone conversations. We don't bother to tell our names if the line is good and we're familiar with the callee. However, radio transmitters are machines and can be replicas of the same mold, where the sender cannot be distinguished by the radiosignal it is emitting. Some louder, some more quiet, but all are using the same voice characteristics, vocabulary, and language! Hence, it becomes essential to effective two-party communication that the sender tells both who is speaking and who it is addressing.

Take a look at the chapter illustration again! Now in our context of that crowded market square where voices cannot be (easily) distinguished and imitation is easy, this might attract a few impostors to business. Hence, authentication of users and messages have to be done explicitly, as early as possible in the communication association, and on a continuous basis during the communication session.

There are several well-grounded reasons for letting a wireless access point be open and nondiscriminating to users, in which case access control is not necessary, though channel security can still be required. A friendly owner simply wants to share the networking resource with others passing by or visiting the neighborhood. Or it can be a wireless community network that is part of a social project providing free internet access to people in the village.

It might also be the case that communication security is taken care of in an end-to-end fashion at the network, transport, or application layer of the communication protocol stack.

There is a wide variety of wireless network technologies in use today. Examples include:

- Personal area networks (PAN), such as Bluetooth

- Local area networks (LAN), such as IEEE 802.11 (Wi-Fi)

- Wide area networks (WAN), such as IEEE 802.16 (WiMax)

- Global area networks (GAN), such as GSM

We have chosen for this chapter to study the wireless LAN (WLAN) security services specified in the IEEE 802.11 standard to illustrate how the challenges of wireless network security are being addressed in practice. This type of wireless network access technology is widespread in our daily lives now. It is built into your laptop computer and your mobile smart phone. It is built into hundreds of millions of stationary and portable consumer electronic devices, and it has become the common method for accessing the internet at universities, airports, city centers, and at home.

8.2 Wireless Local Area Networks

8.2.1 The Standard

The IEEE 802.11 standard of wireless local area networks (WLAN) was first issued in 1997 and has been continuously revised and extended over the years. The current standard is the 2007 version [1].

The Wi-Fi Alliance is an association of manufacturers that certifies WLAN products for interoperability, performance, and conformance to a subset of the 802.11 standard. On the successful side, the technology has been extensively deployed to both professional and private users. Wi-Fi chipsets are included in portable computers, mobile phones, and consumer electronic devices in large numbers (387 million in 2008). There are an estimated 750 000 Wi-Fi access points in 144 countries in 2011 that offer internet access.[2] Wireless Wi-Fi routers are the normal gateway equipment used by ISP customers.

However, the story of the security mechanisms of 802.11 is more of a mixed story, including the large security mechanism amendment 802.11i issued in 2004 [2]. But first we have to know some facts about the network architecture.

8.2.2 The Structure

The structure of 802.11 is defined as follows. A Service Set includes stations (STA) and may include Access Points. There are two structural modes of Basic Service Sets. The Independent Basic Service Set (IBSS), also called the ad hoc mode because all stations communicate directly within each set. Then there is the infrastructure mode, where communication between two STAs is mediated by an Access Point (AP). An AP can be connected to a Distribution System (DS), which interconnects two or more Basic Service Sets, each with one AP. This system is named an Extended Service Set (ESS). The DS can be a fixed LAN running a TCP/IP network. A typical example is a university campus wireless network. Figure 8.1 illustrates an ESS with two Basic Service Sets connected by a DS.

Each WLAN device is associated with a unique IEEE 802 MAC (Media Access Control Layer) address of 48 bits. The AP's MAC address is also used as the basic service set identification (BSSID), which uniquely identifies a BSS. Additionally associated with an AP is the Service Set Identifier (SSID), which is a text string of 0 to 32 octets intended to be a readable name of the network service. An Extended Service Set Identifier (ESSID) is a common string naming an ESS. Thus, an AP will contain both the unique BSSID identifier and a nonunique ESSID.

---

[2]The number is expected to double in 2014 according to the Wi-Fi Alliance.

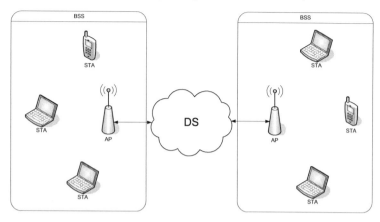

FIGURE 8.1
The IEEE 802.11 infrastructure mode.

MAC Header	Payload	FCS

FIGURE 8.2
An 802.11 frame. The MAC header contains the information needed to forward the payload to its recipient, e.g., the sender and receiver MAC addresses. The frame check sequence (FCS) is a cyclic redundancy check used to detect bit errors in the MAC header and payload. If an error is detected, then the frame is discarded.

8.2.3   Message Types

802.11 messages are called *frames*. There are three different types of frames: *Data* frames, *management* frames, and *control* frames. Data frames are used to transport user data. Management frames are used for signaling, and control frames are used for time-critical signaling. Signaling messages are used to set up, maintain, and terminate a network connection. Figure 8.2 shows a high-level view of an 802.11 frame.

Messages can be either *unicast* or *multicast*. A unicast message has exactly one sender and one receiver. A multicast message has one sender and multiple receivers. In an 802.11 infrastructure network, only the AP is allowed to transmit multicast frames. If a STA wishes to send a multicast message, then first it has to send the message to the AP as a unicast message, and the AP will then forward it to all recipients.

8.3   The 802.11 Security Mechanisms

The security mechanisms have been broken and improved in several rounds. It started out with Wired Equivalent Privacy (WEP), which provides very simplistic means to protect the confidentiality of the data transmitted on the radio link between a STA and the AP. Weaknesses and attacks were soon found once the products arrived on the market, and WEP was completely broken. Then the Temporal Key Integrity Protocol (TKIP) was designed as a WEP fix, and introduced by the Wi-Fi industry as Wi-Fi Protected Access (WPA) while awaiting the slower, but more extensive standardization process of IEEE. The work group of IEEE 802.11i started out in 2001 and the work resulted in two new security protocols, named *Robust Security Network* (RSN), that was included in the 2004 standard. RSN was incorporated in the main standard 802.11 in 2007.

All these security acronyms and configuration possibilities easily confuse the users, which is not a good thing. Normally, a Wi-Fi AP device came out of the box with security disabled by default. One simple noncryptographic solution to access control often provided is to configure and maintain a list of MAC addresses with authorized access in the AP. If the MAC address of the STA is not listed, then access will be denied by the AP. IEEE manages the address space allocation to the manufacturers on a worldwide basis, and a unique MAC address is a assigned by the manufacturer of each network interface card (NIC). Although the MAC address is a globally unique NIC hardware identifier and normally embedded in the hardware, it can often be changed by software techniques. Setting the MAC address by software enables easy and fast MAC address spoofing.

As already stated, WEP turned out to be ill designed and does not provide any protection in practice, as described in the next section. So the next step up will be to configure the device for WPA or WPA2. WPA adds the TKIP protocol to the WEP design, whereas WPA2 is the term for a Wi-Fi Alliance certified implementation of the RSN functionality now included in the IEEE 802.11 standard. Several attacks have shown TKIP to be weak, so the current advice is to configure your 802.11 AP device to WPA2 only. Will the story continue?

8.4   Wired Equivalent Privacy

Wired Equivalent Privacy is the security functionality specified in the original 802.11 standard. Although the security of WEP is badly broken and has been replaced with RSN, it is still found to be commonly used simply because that is what happens to be available. Tews et al. refers to statistics for 490 APs investigated in 2007, where WEP where used by the middle of 2006 in 60% and this dropped to 46% in March 2007 [3]. An investigation[3] for the city of Trondheim in the spring of 2008 estimated about 30% of the total number of mapped AP to be set up with WEP security.

The original security goals of WEP were very modest. The primary goal was "reasonable strong" confidentiality, where data should be protected against eavesdropping over the air link at the strength of Wired Equivalent Privacy. In the 1990s, a major conflict of interests in the international availability of strong crypto surfaced between national security concerns and e-commerce needs. Basically, the interpretation of the multinational Wasenaar agreement on crypto export control back then was that at most 40 bits of encryption key could be accepted. The manufacturers wanted for obvious reasons their WLAN products to be exportable.

The data integrity requirement of WEP said that communication should be protected against message injection and modification, something that was dropped in the corresponding Wi-Fi specification. Moreover, each data-frame (MAC-PDU) should be cryptographically self-synchronizing and self-contained with respect to the parameters needed for processing at the receiver side. This is a significant design decision. And, of course, the security algorithms should be efficient in both hardware and software. Finally, there should be a network access control to protect the use of WLAN communication resources.

The WEP design tried to meet these requirements by a pretty simplistic plan. The cryptographic key management is solved by a single 40 bits key shared among all devices in the BSS, leaving an option for extending the length in the implementations. The key is to be manually shared, updated, and revoked whenever needed. The 802.11 protocol distinguishes between three frametypes: data, control, and management frames. Only the data frames are protected. The integrity is taken care of by appending a checksum of the content excluding the header, and then encrypt this using the shared key.

Sharing a symmetric key among all STA is probably acceptable in a small home network or similar, but it rapidly becomes a problem at larger scales. Envision this scheme put to practice at a university with ten thousands of users, many coming and leaving each year. There is no cryptographic distinction between users, so it is one for all and all for one. How do we manage the cryptoperiods, that is, the validity period of the shared secret key? The key length is certainly too short even to protect against average digital vandals these days. Furthermore, the issue of communication replay has not been tackled, and the management and control frames are not protected at all.

WEP was part of the release of the 802.11 standard in 1997. By 2000, the availability of Wi-Fi devices raised the cryptanalytical interest and some concerns were expressed. By 2001, several serious attacks were published [4] and by 2004 more software utilities for attacks were readily available for public

---

[3]Carried out by Martin Eian.

downloading. By 2007, a 104-bit secret WEP keys could be cryptanalyzed by an active attack in 1 minute [3].

Let us evaluate this design process? On the positive side, it is an *open* standard, which is according the important principle of Kerckhoff. Unfortunately, the design goals were less than clear, the selection of encryption mechanisms was impaired by the ruling international cryptopolicy at the time, and one gets the impression that there was a lack of cryptographic expertise in the committee as well. Certainly, they did not use well-scrutinized crypto-primitives and -protocols, and there was no public review and critique round.

The consequences of securing the WLAN by the WEP plan is that security is optional and not enabled by default, and the security configuration of the devices requires manual user assistance. This state of affairs quickly enabled a new sport called "wardriving," which means roaming about in a vehicle equipped with a standard Wi-Fi enabled portable computer and some special software to hunt down the location and configuration of wireless access points, possibly with the aim to "piggyback" onto an internet connection. By induction, I propose "peacewalking" as the sport where you mount your Wi-Fi enabled pocket computer, put on comfortable shoes for rambling to locate badly configured wireless access points, knock on the owner's door and tell him that you come in peace and ask whether you can be of WLAN configuration assistance. Unfortunately, there seems to be a flaw with that plan as well. Anyway, the label "WEP enabled" now reads "Can use a probabilistic linear time (likely a minute) attack to find the secret key."

8.4.1 RSN with TKIP

One important constraint for the TKIP design was that it must be implementable on existing WEP hardware and protocol encapsulation, but provide additional protection against all known attacks on WEP. Some of the security mechanisms used in TKIP are weak due to this limitation. TKIP is a lot stronger than WEP, but it is gradually falling apart. The first serious vulnerability in TKIP was published by Beck and Tews in 2009 [5] .

We [6] show that the Beck and Tews attack can be used to create an ARP poisoning attack and a cryptographic denial-of-service attack. Moreover, we are able to decrypt DHCP ACK packets, which are 12 times longer than the ARP packets used in Ref [5]. Our method of analysis recovers 596 bytes of keystream that can be used in new attacks on other control messages. TKIP usage should now be avoided, if possible, because of the documented vulnerabilities and the available attack tools. We will therefore focus on the stronger security mechanism, CCMP, in the rest of this chapter.

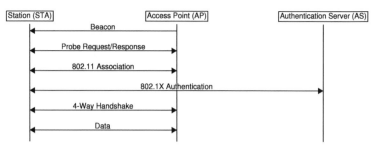

FIGURE 8.3

A high-level view of the 802.11 connection process. A STA first listens for beacons that are periodically broadcasted by the AP. After receiving a beacon, the STA may optionally probe the AP for configuration parameters. The association process includes the exchange of security parameters. The authentication process produces a shared secret key between the STA and AP. The 4-Way Handshake derives the session key.

8.5 RSN with CCMP

8.5.1 Security Services

The strongest cryptographic protocol in the current IEEE 802.11 standard is the Counter Mode with Cipher Block Chaining Message Authentication Code Protocol (CCMP), which uses the advanced encryption standard (AES) to protect protocol messages. The 802.11 standard defines the security services provided by an RSN with CCMP. We will take a closer look at each of these services and describe some technical details behind each of them.

The goal for the security services in an RSN is the same as the original goal for WEP: To provide a level of security equivalent to a wired network. The RSN provides five security services in order to reach this goal: Authentication, data confidentiality, key management, data origin authenticity, and replay detection. Figure 8.3 illustrates the 802.11 connection process for an RSN. The authentication and 4-Way handshake procedures are explained in more detail in the following sections.

8.5.2 Authentication

Authentication means here to establish and verify the identity of the other STA. An RSN supports two authentication methods: preshared key and 802.1X authentication.

The preshared key method is similar to the key management in WEP. Every STA shares a common secret key, and authentication is done by proving knowledge of this key. This method is common in small 802.11 networks, such as a wireless router connected to a home broadband connection. It is often called "WPA2 Personal" in wireless router configuration settings. One weakness of preshared key authentication is that any STA that knows the secret key can impersonate any other STA. In other words, the authentication process can at best only prove that the STA is a member of the group that knows the secret key, it does not identify the individual STA. Another problem with preshared key authentication is that the preshared key is usually based on a password in practice. Poorly chosen passwords can be compromised by offline dictionary attacks, and there are numerous attack tools available that exploit weak RSN passwords.

The second authentication method supported by an RSN is 802.1X [7]. 802.1X is an IEEE standard for port-based admission control, and uses the Extensible Authentication Protocol (EAP) [8] for authentication. 802.1X authentication is often called "WPA2 Enterprise" in wireless router configuration settings. The authentication itself is between the STA and an authentication server (AS). An AS service could be integrated in each AP device, but in large scale deployments, the AS is usually a centralized server for multiple APs. The authentication process includes the generation of a shared secret key between the STA and AS. A new key is generated every time the authentication process is carried out, and this key is *only* known by the authenticating STA and the AS. Once the STA is authenticated, the AS transfers this shared secret key to the AP. The key is later used to negotiate a session key between the STA and AP. Since a new key is generated every time an STA performs the authentication process, and since this key is not known to any of the other STAs, 802.1X authentication is stronger than the preshared key authentication.

The RSN authentication mechanisms supports only point-to-point authentication between the STA and the AS/AP. End-to-end or user-to-user authentication is not supported, and must be provided by upper layers if needed. The authentication is *mutual*, the STA can verify the identity of the AS/AP and vice versa.

8.5.3 Data Confidentiality

An RSN uses encryption to protect the confidentiality of data transmitted over the radio network. Every data frame is encrypted using the AES block cipher in counter mode. The encryption prevents an eavesdropper from reading the contents of the communication. Note that the data confidentiality service is only provided for data frames.[4] Figure 8.4 shows an encrypted CCMP frame.

---

[4]The 802.11w amendment [9], published in 2009, also provides confidentiality for a subset of management frames. Even with this amendment, most management frames and all control frames are still unprotected, something that is bad from a security perspective.

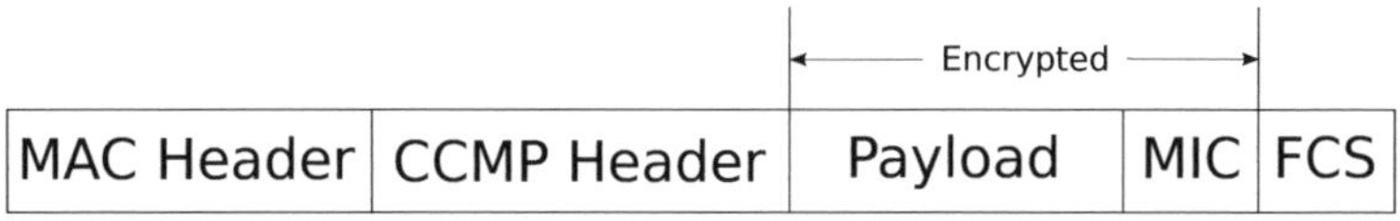

		←——— Encrypted ———→		
MAC Header	CCMP Header	Payload	MIC	FCS

FIGURE 8.4
An encrypted and integrity protected CCMP frame. The CCMP header contains the sequence number. The message integrity code (MIC) protects the integrity of some of the MAC header fields, the CCMP header, and the payload. The payload and MIC are encrypted, while the MAC header, and CCMP header are not.

8.5.4   Key Management

The RSN requires fresh cryptographic keys in order to provide data confidentiality, data origin authenticity, and replay detection services. RSN key management provides fresh keys, and it is implemented through the 4-Way Handshake and the Group Key Handshake.

The 4-Way Handshake is used in the final step of the connection process, as illustrated in Figure 8.3. Once the STA and AP have a shared master key, either preshared or the result of an 802.1X authentication, a temporary session key is generated by the 4-Way Handshake. The session key is derived from the STA MAC address, AP MAC address, STA nonce, AP nonce, and shared master key. In the 4-Way Handshake, the session key is first derived, and the participants then prove knowledge of the key to each other. The resulting session key is called a pairwise transient key (PTK), and it is used to protect unicast data frames.

The Group Handshake is used to distribute group keys from the AP to STAs. The AP generates a group key, and then encrypts the group key with the STA's session key before transmitting it. The group key is called the group transient key (GTK). The GTK is shared by all STAs and the AP, and is used to protect multicast data frames from the AP to the STAs.

8.5.5   Data Origin Authenticity

The data origin authenticity service provides message authentication and message integrity. Message authentication gives the recipient proof that the message sender is legitimate; that is, it proves the identity of the message sender. Message integrity ensures that a message cannot be modified in transit. If a message is modified by an attacker, then the integrity check will fail and the message will be discarded. An RSN with CCMP uses the AES in cipher block chaining message authentication code (CBC-MAC) mode to provide data origin authenticity. It is worth noting that data origin authenticity is only provided for unicast data frames.[5] The reason why this service is not provided for multicast data frames is that the GTK is shared among all the network participants. Even if the STAs are not allowed to transmit multicast messages, there is no guarantee that a malicious STA will obey this rule. Since any STA that knows the GTK can construct a valid multicast message, there is no way to verify the actual sender of the message.

8.5.6 Replay Detection

The replay detection service guarantees that a message is fresh, that is, it is not an old message that was captured by an attacker and replayed later. This service is implemented using a sequence counter that is incremented for every message. The sequence counter is protected by the data origin authenticity service. Similar to the other security services, replay detection only protects data frames.

8.5.7 Summary of Security Services

So far, we have described what the security services of CCMP do. The CCMP functionality provides confidentiality for data frames to protect against eavesdroppers. It also provides message authentication and integrity to protect against attackers trying to impersonate a legitimate STA or AP or modify protocol messages. The authentication service provides access control to the network, so that only legitimate STAs are allowed to connect. Furthermore, the mutual authentication ensures that an attacker cannot set up a rogue AP and trick an STA into believing that this is a legitimate AP. Finally, the security services provide detection of message replay attacks and key management to support the other services.

At this point, one might ask, "What is *not* provided by the security services?" We will take a closer look at this is the next section.

8.6 Assumptions and Vulnerabilities

One of the underlying assumptions of the RSN security is the following:

The destination STA chosen by the transmitter is the correct destination. For example, the Address Resolution Protocol (ARP) and the Internet Control Message Protocol (ICMP) are methods of determining the destination STA MAC address. These are not secure from attacks by other

---

[5]The 802.11w amendment, published in 2009, also provides data origin authenticity for a subset of management frames. Even with this amendment, most management frames and all control frames are still unprotected.

members of the ESS. One of the possible solutions to this problem might be for the STA to send or receive only frames whose final DA or SA are the AP and for the AP to provide a network layer routing function. However, such solutions are outside the scope of this standard. [1]

Most, if not all, 802.11 network implementations use TCP/IP for end-to-end packet switching and data transport. These networks *do* use ARP [10] and ICMP [11] to determine the destination STA, so in practice, the assumption stated above is false! What kind of security failures does this lead to? ARP uses multicast messages to resolve IP addresses. As mentioned in the section on data origin authenticity, *authenticity is not provided for multicast frames*. This means that any STA that is able to connect to the network can send valid multicast messages. If a malicious network member sends a forged ARP message, then it can cause all traffic between the AP and another STA to be redirected to itself. This violates the data confidentiality service, since the malicious STA now can read all traffic to and from the other STA. In other words, the strongest security mechanisms available in 802.11 do *not* protect against active insiders who can eavesdrop by acting as a man-in-the-middle. The failure is caused by a completely unrealistic assumption that is "outside the scope of the standard."

Traditionally, information security is concerned with confidentiality, integrity and availability. The RSN security services provide confidentiality and integrity for data frames. But what about availability? We have repeatedly pointed out that the security services only apply to data frames.[6] Most management and all control frames, used for signaling, are not protected at all! Any attacker can forge those messages at any time to disconnect network users, severely disrupting the network. Attacks against the availability of a network are called denial-of-service (DoS) attacks. Numerous DoS attacks against the 802.11 standard have been published, and we expect more attack methods to be discovered in the future [12, 13, 14].

It should be noted that any wireless network could be subject to jamming, that is, the transmission of noise in the radio frequencies used by the network. Jamming attacks can severely disrupt a wireless network. So why bother with securing the protocols against disruption? There are two main reasons. First, protocol attacks, for example, transmitting a forged management frame, is far more efficient than jamming. By exploiting such vulnerabilities, the attacker can disrupt the network for a long time without transmitting more than a single frame. Such attackers are difficult to locate, and since they are power efficient, they could use battery-powered devices to cause long-term disruption of a network.

A radio jammer has to operate constantly in order to disrupt the network, and is comparatively a lot easier to locate than sending forged messages for a short period of time. The second reason is that attacks using forged protocol frames are much easier to implement in practice.

---

[6]With 802.11w, the security services also apply to a subset of the management frames.

Today, anyone with basic internet searching skills and a small amount of money is able to buy a laptop computer, download attack tools, and carry out significant disruption of 802.11 networks. Jamming attacks usually require specialized hardware,[7] which is illegal to possess in many jurisdictions. Laptops, on the other hand, will probably not be outlawed any time soon.

8.7 Summary

We summarize this chapter with some security recommendations for 802.11 networks. The strongest security available for 802.11 networks is WPA2 with AES-CCMP only. WPA2 by itself is not sufficient to avoid TKIP-related vulnerabilities. If an AP is configured with support for both TKIP and CCMP, then TKIP will be used for all multicast messages [14]. This might compromise network security even for STAs using CCMP for unicast messages.

The 802.1X authentication is stronger than PSK, but if insider attacks are a concern then even this configuration must be complemented by intrusion detection systems able to detect active man-in-the-middle attacks using ARP messages.

For home networks using PSK, the strength of the password used to derive the shared secret key is probably the weakest link. Several attack tools implementing password dictionary attacks against WPA-PSK and WPA2-PSK are available on the internet. A password used as a PSK should ideally be a random string of characters, and as long as possible. If the wireless router is well protected physically, then writing the password on some paper and attach it to the router might actually improve the overall security compared to using a password that is possible to remember. An attacker in a car across the street will not be able to read a note on the wireless router, but can easily break into a network that uses a vulnerable password.

With regards to availability, 802.11 is far more vulnerable than a wired network. This is important to remember if wireless networks are used in safety-critical applications or any other scenarios where availability is a major concern. As an example, consider a wireless surveillance camera used to deter burglars. If a burglar is able to disable the wireless connection from the camera to the server used to store the video recordings, then the camera does not provide its intended function.

---

[7]Certain old 802.11b chipsets had a test mode implementation that could be used as a jamming attack. However, this kind of hardware is not easy to find today, and the software is not readily available.

8.8 Further Reading and Web Sites

The next chapter in this book that describes mobile systems (GSM and UMTS) access security.

Edney and Arbaugh's book is a very readable book about Wi-Fi-protected access [15], though not updated with the recent standards. If you enjoy hands on programming and projects, check out the *Linksys WRT54G Ultimate Hacking* book [16].

The IEEE standards referenced in this chapter can be downloaded from the IEEE website. They contain all the technical details, both for the protocols and the security mechanisms. The standards are available at
http://standards.ieee.org/about/get/

The aircrack-ng tool suite contains implementations of most of the attacks referenced in this chapter. Aircrack-ng is available at
http://aircrack-ng.org/

The BackTrack LiveDVD contains a wide array of security-related tools, including several different wireless network attack tools. Backtrack is available at
http://backtrack-linux.org/

Bibliography

[1] IEEE. *IEEE Std 802.11-2007*. New York, NY, USA, 2007.

[2] IEEE. *IEEE Std 802.11i-2004*. New York, NY, USA, 2004.

[3] Erik Tews, Ralf-Philipp Weinmann, and Andrei Pyshkin. Breaking 104 bit WEP in less than 60 seconds. Cryptology ePrint Archive, Report 2007/120, 2007. http://eprint.iacr.org/.

[4] Scott Fluhrer, Itsik Mantin, and Adi Shamir. Weaknesses in the key scheduling algorithm of RC4. In *Proceedings of the 4th Annual Workshop on Selected Areas of Cryptography*, pages 1–24, 2001.

[5] Erik Tews and Martin Beck. Practical attacks against WEP and WPA. In *WiSec '09: Proceedings of the second ACM Conference on Wireless Network Security*, pages 79–86, New York, NY, USA, 2009. ACM.

[6] F. Halvorsen, O. Haugen, M. Eian, and S. F. Mjølsnes. An improved attack on tkip. *Identity and Privacy in the Internet Age*, LNCS(5838):120–132, 2009.

[7] IEEE. *IEEE Std 802.11X-2004*. New York, NY, USA, 2004.

[8] B. Aboba, L. Blunk, J. Vollbrecht, J. Carlson, and H. Levkowetz. *RFC 3748: Extensible Authentication Protocol (EAP)*, 2004. http://tools.ietf.org/html/rfc3748.

[9] IEEE. *IEEE Std 802.11w-2009*. New York, NY, USA, 2009.

[10] David C. Plummer. *RFC 826: An Ethernet Address Resolution Protocol*, 1982. http://tools.ietf.org/html/rfc826.

[11] J. Postel. *RFC 792: Internet Control Message Protocol*, 1981. http://tools.ietf.org/html/rfc792.

[12] John Bellardo and Stefan Savage. 802.11 denial-of-service attacks: Real vulnerabilities and practical solutions. In *Proceedings of the 12th USENIX Security Symposium*, Berkeley, CA, USA, 2003. USENIX Association.

[13] Bastian Könings, Florian Schaub, Frank Kargl, and Stefan Dietzel. Channel switch and quiet attack: New DoS attacks exploiting the 802.11 standard. In *LCN 2009: Proceedings of the IEEE 34th Conference on Local Computer Networks*, pages 14–21, 2009.

[14] Martin Eian. A practical cryptographic denial of service attack against 802.11i TKIP and CCMP. In Swee-Huay Heng, Rebecca Wright, and Bok-Min Goi, editors, *Cryptology and Network Security*, volume 6467 of *Lecture Notes in Computer Science*, pages 62–75. Springer Berlin / Heidelberg, 2010.

[15] Jon Edney and William A. Arbaugh. *Real 802.11 Security*. Addison Wesley, 2004.

[16] P. Asadoorian and L. Pesce. *Linksys WRT54G Ultimate Hacking*. Syngress Publishing, 2007.

9

Mobile Security

J. A. Audestad

Department of Telematics, NTNU, and Gjøvik University College

CONTENTS

9.1 GSM Security

The Groupe Spécial Mobile (GSM) was established in the autumn of 1982 by the European standardization and policy body CEPT (Conférence Européenne des adminstarsions des Postes et des Télécommunications). In 1988, CEPT was reorganized and the standardization part of CEPT became ETSI (European Telecommunications Standards Institute).

The objective of GSM was to specify a common land mobile system for Europe. This objective was met by the group in 1989 by presenting a complete and unambiguous specification consisting of more than 5000 pages: a group of several hundred people used 7 years to develop this comprehensive standard. The system was put into operation in 1991. The GSM specification is the most successful international standardization effort that has ever taken place.

Already at the third meeting of GSM (in early 1983) it was decided that strong authentication of mobile terminals should be implemented. The reason was that first generation system NMT (Nordic Mobile Telephone) had been introduced in the Netherlands in early 1982. The identity (equal to the telephone number) of these terminals was not protected, and an illegal industry grew up immediately producing cloned mobile terminals. These terminals contained identities and other call parameters that were already assigned to other mobile terminals. The cloned terminals were used in narcotics trafficking and for other criminal purposes. Listening into the radio communications system, everything that was said on the radio connection could be heard, but the actual identities of those who were using the mobile phones remained unknown because the identities of the phones belonged to innocent victims.

At the third meeting of GSM, the Dutch representatives at the meeting had a request that a secure authentication method for mobile terminals was developed for the new European system. This issue was solved at this meeting where a method was proposed that is still used in the newest 3G systems. The basic authentication method is a challenge-response method using a shared key between the network N and the mobile terminal M:

$$N \rightarrow M : \quad r$$
$$N \leftarrow M : \quad s$$

where r is a random number generated by the network, and s (called a signed response) is an encrypted hash (or MAC) of r computed by the mobile terminal:

$$s = H_K(r)$$

where K is the shared key and H is the hash-function. The network, also knowing K, can also calculate s and by comparing this s with the one received from the mobile terminal, the network authenticates the mobile terminal provided that the s is the same in both cases (proves that the mobile terminal knows the secret key K).

The algorithm H_K had to be simple enough for implementation in the mobile terminal. Public key systems were out of the question because of high computational complexity. Therefore, the computationally faster symmetric key method was chosen.

The key K is generated by the mobile user's operator (home network). One copy of K is kept in the SIM of the mobile terminal and one copy is kept in a secure database, called the authentication center, in the home network. This implies that each operator may choose its own authentication algorithm independently of all other operators or may use several authentication algorithms for different groups of mobile terminals, greatly reducing the vulnerability of the system. Furthermore, the authentication center generates all other security information required by the network such as the signed response (s), the encryption key used for message and signaling confidentiality, the message integrity key, and the network authentication token.

In the later development of the GSM and the 3G system, several security features were defined:

- user data and speech confidentiality (in GSM and 3G)

- signaling confidentiality hiding all commands, addresses and other call-related information from eavesdroppers (in GSM and 3G)

- authentication of mobile terminal (in GSM and 3G) as just explained

- authentication of network (in 3G)

- message integrity protection (in 3G)

- user identity confidentiality (in GSM and 3G) where the identity of a user cannot be disclosed by eavesdropping

- user location confidentiality (in GSM and 3G) where the location and, particularly, the change in location of a user cannot be detected by eavesdropping

- user untraceability (in GSM and 3G) where the type of service the user is engaged in cannot be detected by eavesdropping

Some of these security aspects will be discussed next. In order to do so, we need to understand the architecture of GSM/3G.

9.2   3G Architecture

Figure 9.1 shows the general architecture of 3G systems (also called UMTS: Universal Mobile Telecommunications System). Some functions that are not important for security have not been included in the figure. The GPRS architecture of GSM is essentially the same as that of 3G. The major difference is that the RNCs in GPRS are not interconnected in an UTRAN network. This feature is not important for the discussion of the security functions. The GSM architecture without GPRS supports only speech communication and very low-rate data communication. The SGSNs in 3G and GPRS are then replaced by telephone exchanges (MSCs) as explained below. GSM offers authentication of the mobile subscriber and encryption of the speech channel over the radio path. The method used in GSM will be explained in Sections 9.4 and 9.7.

Starting from the left, the 3G system consists of

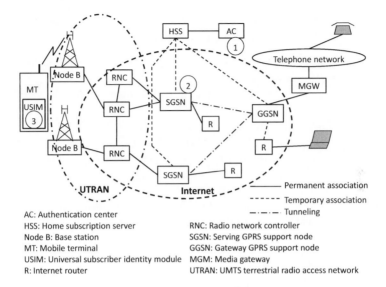

FIGURE 9.1
3G architecture.

- The mobile terminal (MT) (also called user equipment (UE) in some UMTS specifications). While idle, the mobile terminals will monitor a broadcast channel from the Node Bs containing information concerning the identity of the operating cell. The identity of the cell consists of three numbers: network operator identity, location area identity, and node B identity. The mobile terminal will initiate location updating if it receives a new network operator identity (roaming from one network to another) or the same network identity but a new location area identity (roaming within the same network).

- The universal subscriber identity module (USIM) contains identities, user-specific information (e.g., phonebook) and procedures (e.g., IPsec for end-to-end protection), system information, location information, and security parameters and algorithms. In the GSM system, this module was just called SIM – the USIM is thus a SIM with enhanced functionality. The USIM is an application contained in a tamper-resistant hardware device called the universal integrated circuit card – UICC. See specification 3GPP TS 31.102 [1].

- Node B is the base station covering a radio cell. Node B contains radio receivers, transmitters, antennas, and equipment managing the multiple access technique used in the cell.

- RNC is the Radio Network Controller communicating with and managing one or more Node Bs. The RNCs are also internet routers that communicate with one another. The RNCs together with the Node Bs make up the UMTS terrestrial radio access network (UTRAN).

- SGSN is the Serving GPRS Support Node first introduced for GPRS (General Packet Radio Service). The equivalent entity in GSM is the pair of equipment called MSC (Mobile-Services Switching Center – a telephone exchange) and VLR (Visitor Location Register – a database containing subscription data and location information). The SGSN is an internet router also charged with a number of other duties such as managing a number of RNCs, paging for mobile terminals over several RNCs, managing location updating of HSS and GGSN, retrieving identity and other information from other SGSNs when required, and managing security functions. The SGSN is connected to other internet routers. When a new mobile terminal registers itself at the SGSN, the SGSN checks first with the HSS of the mobile terminal if access should be granted. If so, the SGSN receive a security vector from the HSS for authentication. If authentication and registration is successful, the SGSN updates the HSS and the GGSN with the new location data of the mobile terminal.

- GGSN is the Gateway GPRS Support Node. This is the gateway between the internet and the mobile network operator. The GGSN is also an internet router connected to other internet routers. Every 3G operator must have at least one GGSN. The IP number of a mobile terminal is geographically allocated to a GGSN/HSS in the home network of the terminal so that an IP packet addressed to a particular mobile terminal registered in that network is routed to this GGSN. The GGSN possesses location information for the mobile terminals (or retrieves this information from the HSS) so that it knows to which SGSN a particular mobile terminal is connected at any instant. The GGSN uses tunneling to transfer the IP packets to the correct SGSN.

- HSS is the Home Subscription Server of the mobile terminal. The HSS contains all service credentials the mobile terminal has. The HSS also contains location data and manages the retrieval and distribution of security parameters.

- AC is the Authentication Center. This is a secure database calculating all security data the network needs without disclosing any secrets.

- MGW is the media gateway connecting the telephone network, video networks, and so on to the GGSN. In this figure, 3G is an internet extension supporting IP and related protocols (e.g., voice over IP).

The RNCs, the SGSNs, and the GGSNs are part of the internet as shown and are connected to ordinary internet routers as well as to entities within the mobile network.

A call from the internet will be routed to the appropriate GGSN based on the IP number of the mobile terminal. The GGSN will then check if the access is allowed before the IP packet is tunneled to the SGSN to which the mobile terminal is attached using mobile IP tunneling. This connection is normally over several ordinary internet routers. The original IP number will then identify the mobile terminal and the SGSN initiates paging for the mobile terminal over the radio network. Paging for the mobile terminal usually takes place over several RNCs.

An IP packet from the mobile terminal is sent directly into the internet by the SGSN, possibly after having completed an access control check.

Security information is held at three places enumerated 1, 2, and 3 in the figure. In this context, we are not distinguishing between the SGSN and the RNCs – it is easiest to look as these as one site when it comes to information securities. The three sites are

- The authentication center (AC) which produces the authentication vector consisting of the parameters: random number ($RAND$), expected signed response ($XRES$), the session key used for encrypting the message over the radio interface (CK), the message integrity key (IK), and the network authentication token required for network authentication ($AUTN$). A set of vectors (default 5) are sent to the SGSN. The secret information contained by the authentication center consist of the secret key shared with the USIM of the mobile terminal, the algorithm for computing the signed response, the sequence number identifying the vector, and the algorithms for calculating the encryption key, the integrity key and the anonymity key (see below). Note that the algorithm may be publicly known (as is the case for 3G) since the algorithms must be such that it is hard to find, say, the encryption key even if the algorithm and a large amount of cipher-text and corresponding plaintext are known.

- The SGSN/RNC complex contains the set of vectors received from the AC. However, only the random number and the network authentication token are sent on the radio path to the USIM. The other information must be kept in a secure place in the SGSN. The SGSN can verify the signed response from the mobile terminal, encrypt and decrypt messages sent on the radio path, and verify the integrity of received messages.

- The USIM of the mobile terminal contains the same secret information as the AC.

9.3   Extent of Protection

Figure 9.2 shows the parties involved in the security in a 3G system. These are: the mobile terminal (MT), the serving GPRS support node (SGSN), the

communication peer, the home subscription server (HSS), and the authentication center (AC). All communication messages between the MT and the peer
are transited via the SGSN router.

The hierarchy of trust in this model is as follows:

• Trust exists between the HSS and the AC. In this context, the HSS is a
  mediator between the SGSN and the AC.

• Trust also exists, by subscription, between the mobile terminal and the HSS.

• Trust is induced between the SGSN and the HSS during location updating.
  These procedures are not using cryptographic methods to establish trust but
  are based on the assumption that it is difficult for an intruder to impersonate
  as an HSS or an SGSN since a rather secure signaling system exists between
  them (Signaling System No. 7).

• There is no trust initially between the network and the mobile terminal and
  vice versa.

• There is no mutual trust between the mobile terminal and the peer and
  between the peer and the SGSN.

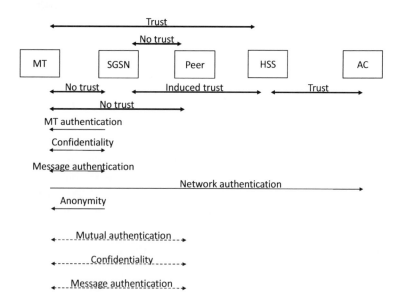

FIGURE 9.2
Security in 3G.

The purpose of the security mechanisms in 3G (and GSM) is to establish
mutual trust between the mobile terminal and the network and to provide
anonymity, confidentiality, and integrity services. Security functions are

- The network (SGSN) authenticates the mobile terminal at the initiation of a communication session.

- Confidentiality using stream cipher is offered on the radio connection to avoid eavesdropping. No protection is provided for the remainder of the connection (see below).

- Two-way message authentication is provided to protect against integrity violation of signaling messages sent over the radio interface.

- The mobile terminal authenticates the network (or, more precisely, the authentication center) at the initiation of a session.

- The SGSN provides anonymity; that is, secure identification of the mobile terminal without disclosing the real identity of the terminal. This function is described in Section 9.8.

Mutual authentication, integrity protection, and confidentiality of the whole communication session between the mobile terminal and the peer are not offered within the context of 3G. This can be implemented directly using IPsec for authentication, confidentiality, and key management.

See the 3GPP recommendations TS 33.102 for the security architecture of 3G systems [TS 33.102].

9.4 Security Functions in the Authentication Center

9.4.1 3G

Figure 9.3 shows the security mechanisms implemented in the authentication center (AC). The authentication center requires a set of parameters and algorithms: five functions (f_1 to f_5), the secret key shared with the mobile terminal (K), a random number generator for generating $RAND$, a sequence number generator (SQN), and the authentication and key management field (AMF). The SQN can be generated using several different algorithms. The AMF field is only loosely defined. This field may consist of zeros only.

The AC then calculates the various security parameters as follows:

- The network authenticator $MAC = f_1(K, RAND, AMF, SQN)$ using the function f_1 and the parameters K, $RAND$, AMF, and SQN.

- The expected signed response from the mobile terminal $XRES = f_2(K, RAND)$.

- The session key for encryption of speech and data packets $CK = f_3(K, RAND)$.

- The integrity key for protecting signaling and related messages $IK = f_4(K, RAND)$.

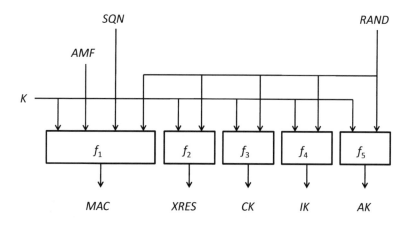

FIGURE 9.3
Security functions in the authentication center.

- The anonymity key $AK = f_5(K, RAND)$.

The functions f_2, f_3, f_4, and f_5 are different functions in the sense that they produce different results even if the input is identical.

The authentication token is then

$$AUTN = (SQN \oplus AK) \mid AMF \mid MAC.$$

$\oplus$ is the XOR operation and $\mid$ is concatenation. The anonymity key AK is only used to mask the sequence number SQN so that it cannot be used for tracing the mobile terminal if the eavesdropper knows the algorithm by which the sequence number is generated. $AUTN$ is sent unencrypted on the radio path. An eavesdropper must not be able to extract any information from it that can be used in a security attack. This is so because the sequence number SQN is different in different $AUTN$ and AK and MAC are pseudorandom numbers.

The authentication vector sent to the SGSN is

$$AV = RAND \mid XRES \mid CK \mid IK \mid AUTN.$$

This is a concatenation of information to be provided to the USIM of the mobile terminal over the radio interface ($RAND$ and $AUTN$), keys to be used by the SGSN for encryption and integrity (CK and IK), and information required for authentication of the USIM ($XRES$). Since the length of all these information elements are system constants, the SGSN can find the various elements in a unique way.

9.4.2 GSM

The GSM system does not support authentication of the access network and message integrity protection. The authentication vector generated by the authentication center and provided to the MSC/SGSN consists then only of the three elements *RAND, XRES,* and *CK.*

9.5 Security Functions in the SGSN/RNC

The security functions in the SGSN/RNC complex are simple. The SGSN receives the authentication vectors from the HSS. A fresh *AUTN* and *RAND,* are sent to the mobile terminal for mutual authentication whenever the terminal accesses the radio system. Each *AUTN, RAND* and other fields of the authentication vector are deleted from memory as soon as they have been used – the same authentication vector is never reused. *AUTN* and *RAND* are sent before the radio path is encrypted since the mobile terminal must receive *RAND* before encryption can take place.

The mobile terminal will compute the signed response *RES* from the random number and send it to the SGSN. The SGSN compares the *RES* received from the mobile terminal with the *XRES* contained in the authentication vector. If they are identical, the mobile terminal is authenticated (proving knowledge of the secret key K).

RNC uses *CK* and *IK* for confidentiality and integrity, respectively. The encryption and integrity functions are described in Section 9.7.

9.6 Security Functions in the Mobile Terminal (USIM)

The security functions in the USIM are shown in Figure 9.4.

The USIM is instantiated for a particular user. The instantiation data related to security are K and the functions f_1, f_2, f_3, f_4, and f_5.

The USIM receives *RAND* and *AUTN* from the SGSN. The USIM may estimate the range in which *SQN* contained in *AUTN* is expected to be found and only accept sequence numbers within a certain range. Certain replay attempts may be discovered in this way.

The USIM computes $AK = f_5(K, RAND)$. *AK* is then used to extract $SQN = AK \oplus (SQN \oplus AK)$ from *AUTN*.

The USIM then computes the expected value of *MAC* from the parameters received from the network and the secret key as follows:

$$XMAC = f_1(K, RAND, AMF, SQN).$$

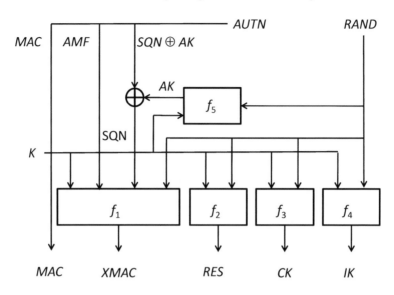

FIGURE 9.4
Security functions in the USIM.

The *MAC* received from the SGSN is then compared with *XMAC*. If they are equal, it proves that the authentication center knows the shared key *K*, and the network is thus authenticated. If they are not equal, the mobile terminal interrupts the call attempt since it is then likely that the access is initiated by someone masquerading as a network.

The USIM calculates $RES = f_2(K, RAND)$ and sends the result to the SGSN. Furthermore, the USIM calculates *CK* and *IK*.

9.7   Encryption and Integrity

9.7.1   Encryption in GSM (A5/1)

The encryption algorithm in GSM is called A5/1. A related but less secure algorithm called A5/2 was implemented for exportation to regions outside Europe.

Although the A5/1 algorithm was kept secret when it was designed in 1987, it has later become publicly known [2]. A5/1 was developed by a very small subgroup of experts of WP3 of the GSM group in order to keep the algorithm secret. The algorithm was then afterward provided by ETSI (the European Telecommunications Standards Institute) to GSM network operators on a licensed basis signing a clause that the algorithm should not be made available to third parties without the permission of ETSI. However, the algorithm became publicly known in 1994 (first published in a magazine for hackers) though it was never admitted by ETSI that the published algorithm was correct.

It is easier to understand how security is implemented in GSM by first taking a look at how the radio access system is composed. This is shown in Figure 9.5. The upper part of the figure shows the general organization of frames and bursts. The access technology is a combination of frequency division multiple access (FDMA) (not shown in the figure) and time division multiple access (TDMA). FDMA implies that different channels are transmitted in nonoverlapping frequency bands. TDMA is used within each such band where the time is split into frames consisting of eight nonoverlapping bursts. Each frame is 4.615 milliseconds long, and each burst has a nominal duration of 577 microseconds. The frame pattern is repetitive so that (say) channel 1 is repeated every 4.615 milliseconds as shown in the figure.

The repetitive sequence of bursts can carry one telephone conversation, or be used for transfer of IP packets, SMS messages, signaling information, and so forth. One base station may support a single frequency with eight bursts equivalent to a capacity of seven telephone channels and one signaling and control channel, or several frequencies with eight telephone channels each if required by traffic demand.

Figure 9.5 also shows the composition of the bursts (direction of transmission is toward the left). The burst consists of

- Tail of 3 bits allowing time for the burst to achieve maximum energy.

- A data field consisting of 57 bits, which is one half of the payload in the burst.

- A stealing set to 1 if this data field is used for other purposes (e.g., to transfer urgent control information or a segment of an SMS message).

- A midamble used for burst synchronization and measurement of the transfer function of the radio channel for adaptive error correction.

- Another stealing flag controlling the usage of the next data field.

- The data field containing the second half of the payload.

- A tail allowing time for the energy of the transmitter to die out.

- A guard time of 8.25 bits ensuring that adjacent bursts do not overlap in time taken into account the maximum allowed instability in clock frequency, Doppler shift caused by fast-moving terminals, and burst misalignment.

The lower part of the figure shows the duplex operation of the channel. The downlink is the direction from the base station to the mobile terminal,

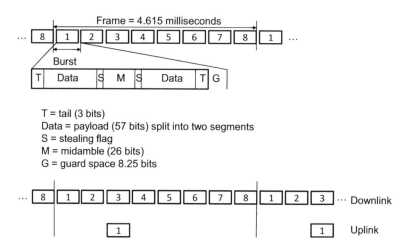

T = tail (3 bits)
Data = payload (57 bits) split into two segments
S = stealing flag
M = midamble (26 bits)
G = guard space 8.25 bits

FIGURE 9.5
Organization of the radio channel in GSM.

and the uplink is the reverse direction. The uplink and downlink use different frequency bands. In the uplink direction, a single burst is allocated to a mobile terminal engaged in a telephone call. For data transmission, the bit rate per channel can be increased by allocating more than one burst to the mobile terminal (GPRS).

Note that the payload per burst consists of only 114 bits.

The frames are organized in multiframes (26 or 51 frames), superframes (26×51 frames), and hyperframes ($2048 \times 51 \times 26$ frames). Multiframes and superframes are required for organizing the communication channels. The hyperframe has a duration of 3 hours, 28 minutes, 53 seconds, and 760 milliseconds and is used as a counter variable in the generation of the key stream. The GSM group decided to use an encryption method where each burst is encrypted individually starting with the same seed in the shift registers (all zeroes). The requirement was then that the keystream generator should be fed with the session key and a counter variable that did not repeat itself in less than 3 hours. A period of 3 hours was chosen because it would be very unlikely that any conversation lasted that long. The result was the hyperframe which allows exact calculation of the counter variable both at the mobile terminal and the base station. 22 bits are required in order to express the maximum hyperframe number in binary form.

Moreover, the same keystream should not be used for encryption of corresponding bursts in the two directions since otherwise it would be easy to derive information from the 2 bit streams that could be used for decrypting them.

The resulting encryption system is shown in Figure 9.6. The encryption key CK (64 bits) and the current hyperframe number HFN is fed to the generator. The generator produces a keystream consisting of 228 bits, where the first 114 bits are used for the downlink and the next 114 bits are used for the uplink. Since *HFN* is incremented by one for each successive frame, the keystream will be different for each frame until the *HFN* repeats itself after about 3 hours and 28 minutes (the duration of the hyperframe).

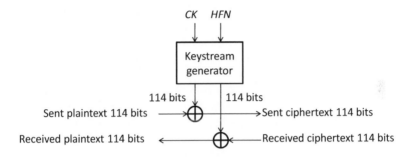

FIGURE 9.6
Stream cipher in GSM.

The actual algorithm is shown in Figure 9.7. This is a good example of keystream generation using a nonlinear combination of linear feedback shift registers.

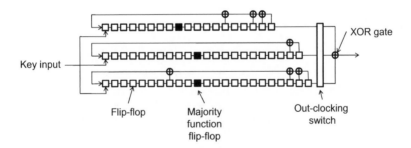

FIGURE 9.7
The A5/1 generator.

The generator consists of three linear feedback shift registers consisting of 19, 22, and 23 stages (flip-flops), respectively. The algorithm is as follows.

First, all flip-flops in the register are set to zero. The *HFN* is concatenated to *CK* producing a string of 86 bits. These bits are then clocked into the three registers (shown as "key input" in the figure) in parallel with the out-clocking switch closed. When all 86 bits have been exhausted, the shift registers are clocked 100 times more as follows. One flip-flop in each register is marked with black in the figure. The content of these flip-flops is the bits used in the majority function. The value of the majority bits is checked before each clocking of the registers. If all three majority bits are equal, then all three registers are clocked; if only two of them are equal, then the two corresponding registers are clocked while the remaining register is not clocked. At the end of this initialization process, the content of the registers will hide the original key and hyperframe number efficiently.

The keystreams are then produced by opening the out-clocking switch and then clocking the registers in parallel using the majority function; that is, only registers belonging to the majority are clocked. The output bits are then added modulo 2. This is done 228 times, producing the keystreams for the two directions.

Note that the majority function is nonlinear. By closer examination of the correlation properties of the keystream, it is found that the algorithm produces a sequence of almost uncorrelated bits and is thus robust against correlation attacks.

It is simple to crack A5/2. A5/1 is a little harder, but this algorithm has also been cracked using various methods.

9.7.2   Encryption in 3G

9.7.2.1   Method

3G is also using stream cipher, but separate keystream generators are applied on the downlink and the uplink. The keystream generator can produce blocks of keystreams from 1 to 2^{16} - 1 bits.

Figure 9.8 shows the encryption method (i.e., the function f_8) used in 3G. Both the mobile terminal and the radio network controller (RNC; see Figure 9.1) are equipped with the same algorithm f_8 producing a keystream segmented into blocks of a certain length. The block length is determined by the organization of the radio channels on the radio interface. The channel structure depends on several parameters such as bit rate, multiple access method, and coding. The block length (the LENGTH parameter) is fed into the function f_8 in order to produce blocks of the required length. The LENGTH parameter consists of 16 bits.

The cipher key CK is 128 bits long. CK is computed using the function f_3 (see Figure 9.4). Note that this key depends on the number $RAND$ so that a new key is computed every time the mobile terminal is authenticated; that is, essentially every time a new radio connection for transfer of control information (e.g., location updating), telephony, data, SMS, MMS, or other information is established.

COUNT-C (the C stands for confidentiality) is a rather complex sequence number among others depending on a hyperframe number in a similar way as in GSM just explained and other sequence numbers related to the mode of transmission. COUNT-C is 32 bits long. The key generator is reset for every frame. COUNT-C ensures that the key stream will be different for all frames over a long period of time so that repetition of the same keystream will be very unlikely. This is the same procedure as in GSM.

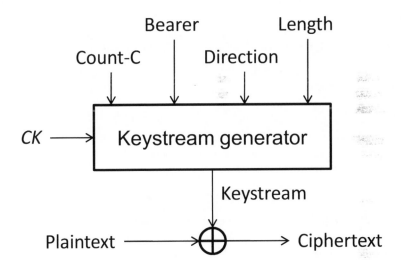

FIGURE 9.8
Encryption.

BEARER contains 5 bits and distinguishes between different channel types (e.g., telephony, packet data, signaling, and short message). The keystream will then be different for different bearer types even if all other parameters are the same. DIRECTION is a 1-bit field with value 0 for the direction from the mobile terminal to the RNC and 1 in the opposite direction. This parameter ensures that the keystream is different in the two directions even if all other parameters are the same. If the keystream were the same (say k) in the two directions, a simple attack would be to add the cipher-texts sent in the two directions producing $c_1 \oplus c_2 = (k \oplus m_1) \oplus (k \oplus m_2) = m_1 \oplus m_2$ and then applying statistics to $(m_1 \oplus m_2)$ to decode the messages partly or completely.

The keystream is finally added modulo 2 (XORed) to the plaintext bit by bit.

Encryption is done by the algorithm outlined below. The algorithm is a block coding algorithm run in a mode that produces a keystream.

9.7.2.2   Keystream Generation Algorithm

The block ciphering algorithm KASUMI is used to produce the keystream. This is a Feistel type algorithm using eight stages. The block length is 64 bits and the encryption key is 128 bits.

For a definition of the algorithm, see 3GPP TS 35.202: KASUMI Specification [3].

9.7.2.3   Initialization of the Keystream Generator

First, the algorithm is initialized. This is done by producing a string C from the input parameters:

$$C = COUNT\text{-}C \mid BEARER \mid DIRECTION \mid 00...0.$$

As usual $\mid$ denotes concatenation. The final field consists of 26 zero bits so that the total length of C is 64 bits. From this string, the initial value A of the algorithm is produced:

$$A = \text{KASUMI}_{CK \oplus KM}(C)$$

where $KM = $ 0x55555555555555555555555555555555 and CK is the cipher key (0x denotes hexadecimal representation). The notation means that KASUMI uses the encryption key $CK \oplus KM$ to produce A.

9.7.2.4   Production of the Keystream

The algorithm is shown in Figure 9.9. KASUMI is used iteratively to generate a number of 64-bit blocks B_i of the key stream. This is done by the following algorithm:

$$B_i = \text{KASUMI}_{CK}(A \oplus (i - 1) \oplus B_{i-1})$$

where i is the block number $1 \leq i \leq n$ where $n = \lceil \text{LENGTH}/64 \rceil$ is the number of blocks that must be generated and $\lceil x \rceil$ is the rounding up operator, for example, $\lceil 128/64 \rceil = \lceil 2 \rceil = 2$ and $\lceil 129/64 \rceil = \lceil 2.015 \rceil = 3$. In the latter case the keystream consists of all 64 bits from both the first and the second block and just the first bit of the third block. Remaining bits are discarded. The keystream is then

$$KEYSTREAM = B_1 \mid B_2 \mid B_3 \ldots \mid B_1^T$$

where B_n^T means that the final block may be truncated if necessary as just explained.

In the first block, we have set $i = 0$ and $B_0 = 0$ so that all blocks are equal. Adding these values to A does not alter A.

This section has demonstrated how external parameters are used in order to make the keystream different for different bearer services, in different frames, and in different directions. This will make attacks on the system more complicated since information extracted for, say, one bearer service cannot be used for attacks on another bearer service.

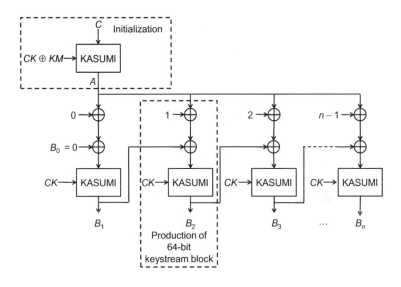

FIGURE 9.9

Keystream generation.

9.7.3 Integrity in 3G

Figure 9.10 shows how the integrity parameter MAC is derived. The MAC is attached to messages sent on the radio path.

The key *IK* (128 bits) is input to the keyed hash-function f_9. COUNT-I (32 bits) is a sequence number similar (but not identical) to that used for encryption, Message is the message to be protected, DIRECTION is a one-bit field that is 0 for messages sent by the mobile terminal and 1 for the opposite direction, and FRESH (32 bits) is a nonce to protect against replay.

The algorithm f_9 uses KASUMI as shown in Figure 9.11.

Using concatenation, the following string of bits is formed:

$$S = COUNT\text{-}I \mid FRESH \mid Message \mid DIRECTION \mid 1 \mid 00\ldots0,$$

where the number of zero bits added at the end of the string makes the string an integer multiple n of 64, and 1 is a single bit with value one. The smallest number of bits in S will then be 65 before the trailing bits are added, so that there will always be at least two blocks. The string is split up into n blocks B_1 through B_n; that is, $S = B_1 \,\hat{}\, B_2 \,\hat{}\, \ldots \,\hat{}\, B_n$. Each block is then fed into KASUMI together with the result from the previous block, thereby producing the output (setting $A_0 = 0$):

$$A_i = \text{KASUMI}_{IK}(B_i \oplus A_{i-1})$$

KM is the same fixed key that is defined in Subsection 9.7.2. Finally, the 32 leftmost bits of C is used as the 32-bit MAC.

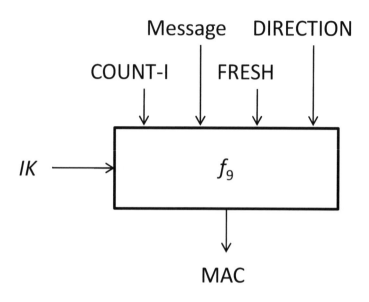

FIGURE 9.10
Derivation of message integrity code (MAC).

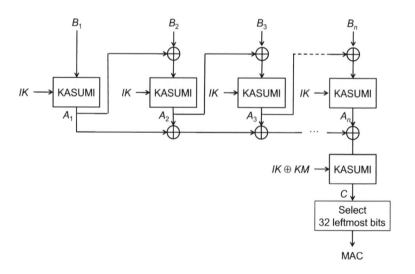

FIGURE 9.11
Integrity algorithm in 3G.

9.8 Anonymity

Anonymity on the radio path is achieved by allocating a temporary mobile subscriber identity, TMSI, to the mobile terminal as explained below.

The mobile terminal is uniquely identified globally by its international mobile subscriber identity (IMSI). The IMSI consists of three fields: (1) a country code identifying the country in which the mobile terminal is registered, (2) a network code identifying a particular mobile network operator of that country, and (3) a unique identity of the mobile terminal within the network. The country code is allocated by the International Telecommunications Union (ITU), the network code is allocated by a national regulatory or standards body, and the unique identity within the network is allocated by the network operator. The IMSI is public knowledge; that is, knowing the IMSI, it is possible to identify the subscriber associated with that IMSI.

Note that encryption can only take place after the authentication has been completed – or, in other words, after the mobile terminal has received *RAND*. On the other hand, authentication cannot take place before the identity of the mobile terminal is known and associated with the authentication vector to be used. Therefore, the initial messages containing the identity must be sent on the radio path before encryption is turned on. If the temporary identity used in this message can be kept secret, it is difficult for anyone to trace the mobile terminal. The protection of the temporary identity consists of two security precautions:

- not disclosing the real identity of the mobile terminal while it is roaming into an area controlled by a new SGSN, and

- frequently changing the TMSI allocated to a mobile terminal in such a way that even if the TMSI is known from a previous transaction, it is difficult to predict the next value of the TMSI.

The way in which this is achieved is shown in Figure 9.12. The enumeration of messages indicates the order in which messages and actions are sent or executed. When the mobile terminal roams from the area controlled by one SGSN to the area controlled by another SGSN, the location information contained in the HSS and the GGSN must be updated. Otherwise, telephone calls or data packets cannot be routed to the terminal. The mobile terminal knows when updating must take place from information broadcast to all mobile terminals in the area controlled by each SGSN.

The identity of the mobile terminal used in the location updating message is the TMSI allocated by the previous SGSN together with the identity of that SGSN (message 1 in the figure). The new SGSN can then retrieve the IMSI and remaining authentication vectors (AVs) from the previous SGSN (messages 2 and 3). The new SGSN then authenticates the mobile terminal (message exchange 4), and if the authentication is successful, the SGSN initiates

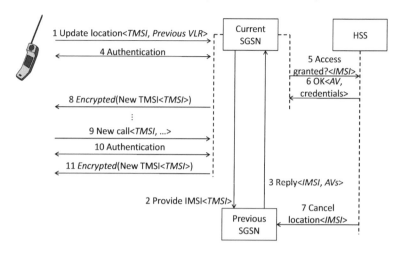

FIGURE 9.12
Location updating, connection setup, and anonymity.

the location updating procedure with the HSS of the mobile terminal (message 5). If the access is granted, the HSS returns a set of authentication vectors for future use and subscription credentials allocated to the mobile terminal (message 6). The HSS also send a cancel location message to the previous SGSN so that the mobile terminal can be deleted from the database (message 7).

The SGSN now initiates the encryption procedure, allocates a new TMSI, and sends it in encrypted form to the terminal (message 8). This TMSI is used the next time the terminal accesses the SGSN (message 9). If authentication is successful (message 10), the SGSN allocates a new TMSI and provides it in encrypted form to the terminal (message 11).

The IMSI is then only used the first time the mobile terminal accesses the system, if the TMSI has been lost from the memory in the USIM, or if the SGSN does not support the procedure just described. The latter may apply when the mobile terminal roams to a different network.

Note that the GSM/3G specifications allow the network to force the mobile terminal to send the IMSI instead of TMSI by returning an error message to the mobile terminal in the location updating sequence.

9.9 Example: Anonymous Roaming in a Mobile Network

9.9.1 Procedure

This section is based on a proposal for anonymous roaming and undeniable billing in mobile systems by J. Zhou and K. Y. Lam [4]. Below, only the anonymous roaming procedure is described. The procedure may be applicable in both public (3G) and private (WiFi) mobile networks.

The principle is shown in Figure 9.13. Initially, roaming agreements exist between the home network (HN) of the mobile terminal and foreign networks (FNs) that the mobile terminal is allowed to access. The mobile terminal registers in a foreign network by sending a location update message to the FN, which allocates a temporary identity to the mobile terminal and forwards this identity and the update message to the home network of the mobile terminal. The home network then allocates a session key to be used between the mobile terminal and the foreign network. The details are as follows.

Notation:

- K_{pubH} is the public key of the home network.

- K_{MH} is the permanent secret key shared between the mobile terminal and the home network. The key is allocated by the home network at subscription.

- K_{FH} is the permanent secret key shared between the foreign network and the home network allocated as part of the roaming agreement.

- K_{MF} is a temporary session key allocated by the home network for the encryption of the communication between the mobile terminal and the foreign network.

- I_{MH} is the permanent identity of the mobile terminal in the home network.

- I_M is a temporary identity assigned to the mobile terminal by the foreign network during call setup.

- I_H is the permanent identity of the home network.

- I_F is the permanent identity of the foreign network.

- S_H is the signature key of the home network.

- S_M and V_M are signature key and verification key, respectively, assigned to the mobile terminal for use in the foreign network. These keys are used to sign messages and verify the signature of messages exchanged between the mobile terminal and the foreign network.

- T_s is the expiry time of the signature and verification keys.

- N_M and N_F are nonces generated by the mobile terminal and the foreign network, respectively.

- $E_x(y)$ denotes encryption of y using key x.

- $S_x(y)$ denotes the signature of x on y.

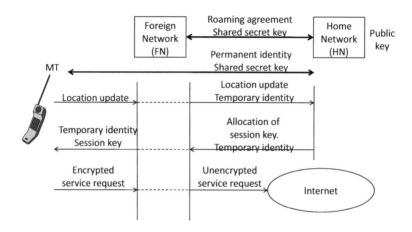

FIGURE 9.13
Anonymous roaming.

The location updating protocol is then:

- MT to FN: I_H, N_M, $E_{KpubH}(N_M, I_{MH}, K_{MH}, I_F)$

- FN to HN: N_F, $E_{KpubH}(N_M, I_{MH}, K_{MH}, I_F)$, $E_{KFH}(I_M, H(I_M, N_M))$

- HN to FN: $E_{KFH}(N_F, I_M, K_{MF}, V_M, T_s)$, $S_H[I_M, I_F, V_M, T_s]$, $E_{KMH}(N_M, I_M, K_{MF}, S_M, T_s)$

- FN to MT: $E_{KMH}(N_M, I_M, K_{MF}, S_M, T_s)$, $H(K_{MF}, N_M)$

The mobile terminal sends the identity I_H of the home network in the location update message. If there is a roaming agreement with the home network, the request is accepted. If not, the request is either rejected or a roaming association is established with the home network using, for example, IPsec for exchange of secret keys. This association must be established before the procedure can continue. This is not shown in the figure and in the message exchange above.

The message also contains a nonce N_M generated by the mobile terminal to prevent replay and a message part destined for the home network $E_{KpubH}(N_M, I_{MH}, K_{MH}, I_F)$. This message part is encrypted by the public key of the home network and contains the nonce, the identity of the mobile terminal in the home network I_{MH}, the secret key shared between the home network and the mobile terminal K_{MH}, and the identity of the foreign network I_F.

The foreign network forwards the message part destined for the home network $E_{KpubH}(N_M, I_{MH}, K_{MH}, I_F)$ plus a nonce N_F and a temporary identity assigned to the mobile terminal I_M to the home network. The temporary identity I_M and a hash over the temporary identity and the nonce produced by the mobile terminal are encrypted by the secret key shared with the home network K_{FH}. Encryption using the shared key K_{FH} together with a valid hash authenticates the foreign network. The home network can verify this message digest from I_F contained in the same message and the nonce N_M received in the message $E_{KpubH}(N_M, I_{MH}, K_{MH}, I_F)$.

When the home network receives the message $E_{KpubH}(N_M, I_{MH}, K_{MH}, I_F)$ from the mobile terminal, the home network will check that the identity and the secret key K_{MH} contained in the message are contained in the subscription database of the home network and that they belong to the same mobile terminal. If they do, the terminal is authenticated. If not, the location request is rejected. The inclusion of both identity and secret key protects against a particular type of denial-of-service attack where an intruder attempts to update the home network with false location information.

The home network allocates the key K_{MF} that the foreign network and the mobile terminal can use for encryption and a signature/signature verification pair $(S_M$ and $V_M)$, allowing the mobile terminal to sign messages sent to the foreign network. An expiry time for these keys (T_s) is also assigned.

The home network then returns two message parts $E_{KFH}(N_F, I_M, K_{MF}, V_M, T_s)$ and $E_{KMH}(N_M, I_M, K_{MF}, S_M, T_s)$, where one part is destined to the foreign network – encrypted by the shared key between the two networks – and the other part is destined to the mobile terminal and encrypted using the shared key between the home network and the mobile terminal. The foreign network receives the nonce sent in the previous message (this prevents replay), the temporary identity of the mobile terminal allocated by the foreign network (I_M), the encryption key to be used between the mobile terminal and the foreign network (K_{MF}), the signature verification key V_M, and the expiry time of the signature T_s.

The message part destined to the mobile terminal contains the nonce generated by the mobile terminal N_M, the temporary identity to be used on subsequent messages I_M, the encryption key to be used between the mobile terminal and the foreign network K_{MF}, the signature key S_M, and the expiry time of the signature key T_s.

The message to the foreign network also contains a signature over the temporary mobile identity, the identity of the foreign network, the signature verification key, and the expiry time for the key $S_H[I_M, I_F, V_M, T_e]$. The signature binds together information in such a way that the signature can be used for nonrepudiation: the home network admits to have seen the temporary identity and that it knows the signature verification key allocated to the mobile terminal.

Finally, the foreign network forwards the information $E_{KMH}(N_M,\ I_M,\ K_{MF},\ S_M,\ T_s)$ from the home network together with a message digest $H(K_{MF},\ N_M)$ over the session key between the mobile terminal and the foreign network and the nonce generated by the mobile terminal in the initial message.

Subsequent messages are then formatted as follows:

$$\text{MT to FH} : I_F, E_{KMF}(m), S_M[I_F, N, m],$$

where m is the message and N is a nonce generated by the mobile terminal for preventing replay. The foreign network can then forward the message unencrypted into the internet. However, the terminal may have established a security association with the peer entity in the internet so that the message is sent in encrypted form between the mobile terminal and the peer using tunneling in the foreign network.

9.9.2 Information Stored

The mobile terminal stores the following information:

- permanent mobile terminal identity in home network, I_{FM}
- temporary mobile terminal identity in foreign network, I_M
- identity of home network, I_H
- identity of foreign network, I_F
- public key of home network, K_{pubH}
- permanent key shared with home network, K_{MH}
- temporary mobile terminal key shared with foreign network, K_{MF}
- signature key, S_M

  The home network stores:

- permanent mobile terminal identity in home network, I_{FM}
- identity of foreign network, I_F
- public/private key pair of home network, K_{pubH}/K_{privH}
- permanent key shared with mobile terminal, K_{MH}
- permanent key shared with foreign network, K_{FH}

  The foreign network stores:

- temporary mobile terminal identity in foreign network, I_M

- identity of home network, I_H

- temporary mobile terminal key shared with mobile terminal, K_{MF}

- permanent key shared with home network, K_{FH}

- signature verification key, V_M

9.9.3 Prevention of Intrusion

9.9.3.1 The Mobile Terminal Is an Impostor

In this case the mobile terminal does not know the combination of shared key K_{MH} and identity I_{MH} and can, therefore, not compose a valid message for the home network in the first step of the protocol. The home network will then reject the attempt and the foreign network will terminate the procedure.

9.9.3.2 Both the Mobile Terminal and the Home Network Are Impostors

In this case the home network cannot construct a valid message in step 3 of the procedure because the home network does not know the shared key K_{FH} between a genuine home network and the foreign network. The nonce N_F ensures that this message cannot be a replay of a message the foreign network has received previously from a genuine home network.

9.9.3.3 The Foreign Network Is an Impostor

In this case the foreign network cannot construct a valid $E_{KFH}(I_M, H(I_M, N_M))$ in step 2 since it does not know the key K_{FH}, and will, for the same reason, not be able to decode $E_{KFH}(N_F, I_M, K_{MF}, V_M, T_s)$ in step 3 and hence construct $H(K_{MF}, N_M)$ in step 4.

9.10 Using GSM/3G Terminals as Authentication Devices

9.10.1 Architecture

Two methods that are used in the mobile network are explained next ([5] and [6]) for authenticating a browser in a personal computer. The general configuration is as shown in Figure 9.14.

The system consists of the following components:

- The user terminal (or browser).

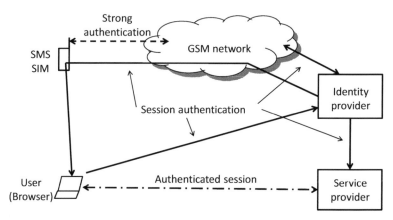

FIGURE 9.14
Session authentication using GSM/3G.

- The service provider offering the service requested by the user.

- An identity provider also acting as an authentication server. The user identity and the user credentials are stored in this server.

- A mobile phone belonging to the user.

- The GSM/3G network.

The identity and credentials of the mobile phone (or rather the SIM or USIM it contains) are managed by this network. The GSM/3G network is also responsible for authentication and encryption over the radio path and supports SMS services. The network also contains naming databases (HLR or HSS; see Section 9.2) and offers identity provider functionality binding several naming elements such as international identity (IMSI), IP number, telephone number, authentication vectors, and possibly other security information.

Several relationships are shown in the figure:

- The SIM/USIM is strongly authenticated by the GSM/3G network. This authentication takes place whenever the mobile terminal accesses the radio network, for example, every time the terminal sends or receives an SMS.

- The session to be authenticated is between the user terminal and the service provider.

- Several other relationships are shown that may be involved when the service provider authenticates the user terminal. These are between the browser and the identity provider, the identity provider and the SIM/USIM, between the service provider and the identity provider, and between the identity provider and the GSM/3G network.

The service provider, the identity provider, and the GSM/3G network are all members of the Liberty Alliance [7]

The two cases explained below are

- Authentication using one time password.

- Authentication using the extensible authentication protocol (EAP).

The initiation starts when the browser accesses the service provider. In both cases, the service provider returns a diversion command requesting the browser to access the identity provider.

9.10.2   One Time Password

The procedure is shown in Figure 9.15. The procedure is as follows. The browser starts the initiation procedure toward the service provider (access request and redirect messages). The browser then requests the identity provider to establish the connection on behalf of the browser. The identity provider checks the identity and credentials of the browser (user) before sending a one time password to the mobile terminal of the user over the GSM network. The mobile telephone number of the user is information that the service provider knows or can obtain by federation (e.g., using the procedures of the Liberty Alliance). The identity provider uses the telephone number in the SMS message.

Before the GSM/3G network sends the onetime password to the mobile terminal on SMS, the network authenticates the mobile terminal (SIM/USIM) and invokes encryption on the radio path as explained earlier in this chapter. The one time password is then provided to the browser either automatically (using Bluetooth or the bus interface) or manually (which is the most common procedure). The browser returns the one time password to the identity provider that, if the password is accepted, sends an accepted message to the service provider so that the user can be invited to continue the logon procedure.

Security in this system is based on the strong authentication and the encryption mechanisms of GSM/3G and on the secure binding of username, telephone number, and IMSI over several administrative domains which comprise at least the service provider, the mobile operator, and the username manager of the GSM network (HLR) or the 3G.network (HSS).

9.10.3   The Extensible Authentication Protocol (EAP)

This section describes the most important issues related to EAP. The basic protocol is shown in Figure 9.16. The browser is supplied with additional functions (e.g., using Java applets) for communicating with the SIM and with the identity provider. The SIM is also extended by a separate (say) Java applet to support the particular procedure implemented across the radio interface.

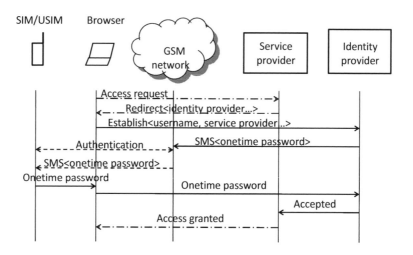

FIGURE 9.15
Authentication of browser using one time password over SMS.

EAP-SIM is defined in RFC 4186 [8].

The SIM may be connected to the browser via Bluetooth, may be installed in the computer, or may be connected via the universal serial bus interface. There is also a variant where SMS may be used for the communication with the SIM. In the figure, connection via Bluetooth is shown.

The identity provider contains the required functionality for supporting the EAP procedures and the communication with the mobile network. The purpose of the procedure is to achieve strong authentication between the browser and the service provider.

The procedure starts as usual with the browser requesting access to a service provider. The service provider redirects the browser to logon with an identity provider supporting the EAP procedure. The authentication provider is also connected to the HLR/AC of the GSM network for management of retrieval of GSM authentication vectors (AVs), that is, triplets consisting of *RAND*, *XRES*, and *CK* in the GSM system (see Subsection 9.4.2). The mobile network, the identity provider, and the service provider may have *a priori* trust in one another via Liberty Alliance. Interworking units may be installed in order to translate between protocols designed in accordance with different standards. One example is interworking between the protocols used in the internet and the protocols used for signaling in the mobile network. The interworking units are not shown in the figure.

After having logged on to the identity provider, the identity provider will request the browser to provide the identity required for authentication (the EAP-request/identity message). The identity is, in this case, the international mobile subscriber identity (IMSI) of the SIM. The Java applet obtains

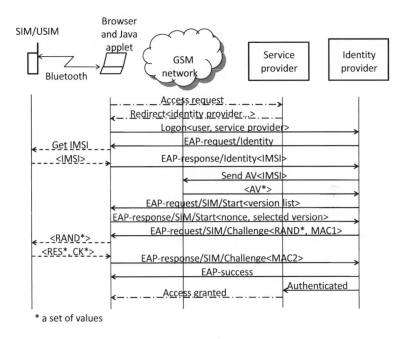

FIGURE 9.16
EAP-SIM authentication.

the IMSI from the SIM and returns it to the identity provider in the EAP-response/identity message. Note that the IMSI is sent in encrypted form over the Bluetooth connection. The IMSI identifies uniquely the home network of the SIM so that the identity provider can obtain a set of authentication vectors (AVs) from the GSM network.

Having received the AVs, the identity provider sends the EAP-request/SIM/Start message to the Java applet, indicating supported versions of the protocol (which may happen to be just one). The Java applet responds by the EAP-response/SIM/Start message containing the chosen version and a nonce. The identity provider then sends the EAP-request/SIM/Challenge message containing 2 or 3 random numbers ($RAND^*$) and MAC1 calculated over the message plus the nonce received in the previous message. The $RAND^*$ are sent to the SIM that calculates and returns the set of responses RES^* (one for each $RAND$ in $RAND^*$) and the set of corresponding confidentiality keys CK^*. The notation x^* is used to indicate a set of variables x with different values (the set consists of 2 or 3 elements in the current version of EAP).

The key the identity manager uses to produce MAC1 is derived from several information elements, including the CK^* received from the GSM network in the AVs; see the RFC for details. The Java applet in the browser can then authenticate the identity provider by calculating the MAC1 using the RES^*

received from the SIM. If the calculated MAC1 and the received MAC1 are identical, the network is authenticated.

If the network is authenticated, the Java applet returns the message EAP-response/SIM/Challenge containing a MAC2 calculated over the message plus the RES^*. The key is derived from the same algorithm as is used for producing MAC1, but the keys used are different. MAC2 then authenticates the SIM since the SIM has calculated the correct set of keys (CK^*) and the correct set of responses (RES^*). This ends the mutual authentication procedure.

The identity provider now terminates the procedure with the applet and informs the service provider that authentication has been successfully completed. The identity provider then grants access to the browser.

9.11 Further Reading

The first book on the GSM system was published by Michel Mouly and Marie-Bernadette Pautet [9]. This is still the most comprehensive text published on the GSM system. A more available book containing almost the same level of detail is *An Introduction to GSM* by S. Redl et al.[10]. A recent book on UMTS is [11].

The book by Valteri Niemi and Kaisa Nyberg [12] gives a comprehensive overview of the security solutions and specifications of UMTS (3G system) both providing the theoretic background of the methods and the practical design principles. Though attempts were made to keep the encryption algorithm of GSM (A5/1) secret, it became publicly known already in 1994. The algorithm is described in Wikipedia [2].

All detailed specifications of UMTS security are contained in the 33-series (Security aspects) [13] and the 35-series (Security algorithms) [14] of 3GPP (3rd Generation Partnership Project). These recommendations are continuously updated so that the latest (and also earlier) versions can be found by following the URL links in the references.

The particular anonymity protocol for mobile access described in Section 10.9 is based on a method proposed in Chapter 7 of the book on non-repudiation by Jianyuing Zhou [4]. The method can be implemented on any local area or public mobile system. The procedures are independent of the access technology of these systems.

Several methods using the SIM or the USIM as an authentication token is described in [5] and [6]. One of the particular algorithms (EAP) is specified in [8] and [15] for GSM and 3G, repectively. These protocols may use Liberty Alliance for federation of identities; see [7].

Bibliography

[1] Characteristics of the Universal Subscriber Identity Module (USIM) application. http://www.3gpp.org/ftp/Specs/html-info/31102.htm.

[2] A5/1. http://en.wikipedia.org/wiki/A5/1.

[3] 3GPP TS 35.202. Specification of the 3GPP confidentiality and integrity algorithms; Document 2: Kasumi specification. http://www.3gpp.org/ftp/specs/html-info/35202.htm.

[4] J. Zhou. *Non-repudiation in Electronic Commerce*. Artech House, 2001.

[5] Thanh van Do. Identity management. *Telektronikk*, 3(4), 2007.

[6] Thanh van Do, T. Jønvik, and I. Jørstad. Enhancing internet service security using GSM SIM authentication. In *Proceedings of IEEE GLOBE-COM*. IEEE, 2006.

[7] Liberty alliance project. http://www.projectliberty.org.

[8] Extensible authentication protocol method for global system for mobile communications (gsm) subscriber identity modules (eap-sim). http://www.ietf.org/rfc/rfc4186.txt.

[9] M. Mouly and M.B. Pautet. *The GSM System for Mobile Communications*. Telecom Publishing, 1992.

[10] S. M. Redl, M. K. Weber, and M. W. Oliphant. *An Introduction to GSM*. Artech House, 1995.

[11] R. Kreher and T. Ruedebusch. *UMTS Signaling: UMTS Interfaces, Protocols, Message Flows and Procedures Analyzed and Explained*. Wiley, 2007.

[12] V. Niemi and K. Nyberg. *UMTS Security*. Wiley & Sons, 2003.

[13] 3gpp specifications, security aspects. http://www.3gpp.org/ftp/Specs/html-info/33-series.htm.

[14] 3gpp specifications, security algorithms. http://www.3gpp.org/ftp/Specs/html-info/35-series.htm.

[15] Extensible authentication protocol method for 3rd generation authentication and key agreement (eap-aka). http://www.ietf.org/rfc/rfc4187.txt.

10

A Lightweight Approach to Secure Software Engineering

M. G. Jaatun, J. Jensen, P. H. Meland and I. A. Tøndel
SINTEF ICT

CONTENTS

10.1 Introduction

Secure software engineering[1] is much more than developing critical software. History has shown us that software bugs and design flaws also represent exploitable security vulnerabilities in seemingly innocuous applications such as web browsers and PDF document viewers. This implies that there is a need for a well-balanced amount of security awareness in all software development projects right from the beginning.

Most software developers are not primarily interested in (or knowledgeable about) security; for decades, the focus has been on implementing as much functionality as possible before the deadline, and then patch whatever bugs there may be when it's time for the next release or hotfix. However, it is slowly beginning to dawn on the software engineering community that security is important also for software whose primary function is not related to security.

There are clear indications that significant cost savings and other advantages are achieved when security analysis and secure engineering practices are introduced early in the development cycle, and that the number of serious security defects can be significantly reduced with a minimum of extra costs. However, having a clear security focus is not easy, and today there are very few people that master both the art of software and security engineering. There is thus a need for closer collaboration and knowledge transfer between the two factions.

SODA is an approach to inject secure software engineering practices into existing software development processes. The SODA target group is the "ordinary" developer, who is not primarily interested in (or knowledgeable about) security, but must focus on designing/implementing as much functionality as possible before the deadline is passed and/or the budget is exhausted.

SODA is based on the following assumptions:

1. A developer will not try to learn or memorize security knowledge prior to starting the development.

2. There should be no significant change in the way developers work.

3. There must be good tool support that enhances security during development, preferably integrated into the current development tools.

In the following sections, we will present how this philosophy is reflected in approaches for asset identification, security requirements elicitation, security design, and security testing. As can be seen from the grayed-out parts of Figure 10.1, we will not cover secure coding, deployment or monitoring, but these topics are covered by several easily accessible books (e.g., [5, 6]).

---

[1] This chapter is primarily based on results from the SODA project, as documented in a series of papers [1, 2, 3, 4]. For more detailed references and background, please refer to these original papers.

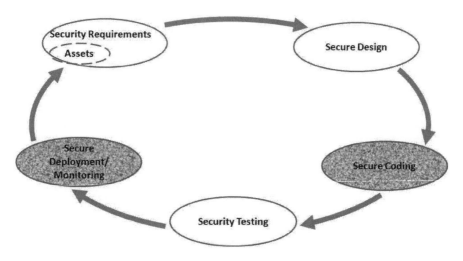

FIGURE 10.1
The main phases of the SODA approach to secure software engineering.

10.2   Assets

10.2.1   Asset Identification

The concept of "assets" is central to the very idea of security – we need security because we have something that needs protection. This "something" is what we collectively refer to as our assets. Thus, asset identification is a crucial component of the requirements phase – specifically, security requirements are primarily needed in order to specify what we need to do to protect our assets, and this will obviously be impossible to do properly unless we know what these assets are. To highlight the importance of the asset identification phase, we detail it here first, separately from the main security requirements phase, which is described in Section 10.3.

The goal of the method we describe in the following is to discover all the assets that are relevant for the system being developed, and facilitate a prioritization process in order to identify which assets have a higher (or lower) priority with respect to security. Strictly speaking, our primary concern is to identify the assets that are *most important* – if we overlook assets that *don't* need protection, we can still sleep at night.

Before asset identification takes place, the main security objectives of the software to be developed should be identified. By security objectives, we mean high-level security requirements or goals identified by customers, and any security requirements coming from standards, policies, or legislation. The results of asset identification should be used as a basis for identifying threats, where attack trees or similar are created based on the most important assets identified. Security requirements are then elicited based on threat analysis.

10.2.2 Asset Identification in Practice

The SODA asset identification method helps to establish an overview of the assets of a system and their different requirements for protection. This information makes it easer to prioritize security requirements (either informally, or as part of a formal risk analysis process) later on. We suggest to look at the value of the assets both from the user's and the system owner's point of view, but also from the view of an attacker. To identify assets, one can use functional requirements of the system in combination with brainstorming techniques. For each asset, one should then make a judgment regarding the different stakeholders' priority of the confidentiality, integrity, and availability of this asset. When assigning priorities, one should use predefined categories, for instance, high, medium, and low.

When an asset identification has been performed, you should have an overview of which properties of what assets are most important to protect. This information should be used to prioritize requirements and to identify where to focus the effort when it comes to identifying more detailed requirements.

10.2.2.1 Key Contributors

The asset identification process should be performed by a diverse group composed of

- The developer's requirements team

- Other developers (not part of the core requirements team)

- Security experts (if available – security expertise will be a bonus in this process, but our method is designed to also work without it)

- Customers and/or end-users (if applicable)

Generally, anyone that could contribute with ideas on assets may contribute, but the total group should not exceed eight persons (plus facilitator). We assume that the participants either are familiar with the system before the asset identification starts, or that a short briefing on the main system characteristics is given as part of the introduction.

This method can be used in any project. For large projects, one may however need to use more coarse assets than in smaller projects, or execute the method several times on a subset of the application domain.

10.2.2.2 Step 1: Brainstorming

Our brainstorming technique may be considered an amalgamation of traditional brainstorming and "brainwriting." The purpose of brainstorming is generally to generate ideas on a given topic; in our case, the topic is restricted to answering the question "what are our assets." By using this technique, one is able to involve different types of persons in a creative idea-generating process, without putting too many restrictions on the end result.

As preparation for the brainstorming session, sticky notes[2] and felt-tip pens/markers must be made available for all participants, and somewhere to put the sticky notes during brainstorming must be provided (e.g., a large sheet of paper to put on the wall).

1. Present the process and the rules to follow while brainstorming:

   - Everybody shall participate
   - No discussion/criticism during the brainstorming
   - One should build on ideas presented by others

2. Give out sticky notes and markers to all participants.

3. Decide on a time limit for individual brainstorming (e.g., 5 minutes).

4. Formulate a question on which to brainstorm (e.g., "What are our assets?"), write it down and place it somewhere visible to everybody in the group.

5. All participants write down their ideas – one idea in large letters per note (the note should be readable from a distance of several meters).

6. When time is up, everybody presents their ideas to the group by placing their notes on, for instance, a piece of paper on the wall. Often it is advantageous to do this in a structured way: Everybody takes turns presenting their ideas. One is only allowed to present a limited number of ideas at the time, and the remaining ideas must wait until the next round.

7. Group the ideas and eliminate duplicates. Everybody should participate in this step.

8. Document the result, for example, with a camera.

Note that although Wilson [7] warns that a trained facilitator is necessary to ensure success of a brainstorming session, we have found that even with just "normal" participants (and appointing one facilitator), there is tangible value to be had from brainstorming. Furthermore, this is also to a great extent *learning by doing* – if a development organization frequently employs brainstorming in relevant situations, the individual participants will in time become reasonably confident (if not to say skilled) as facilitators.

---

[2] *For example, Post-it Notes*®, a registered trademark of 3M (http://www.3m.com). We have found that generic-brand sticky notes also work. (We are also aware that there are various computer-based brainstorming tools that may be used instead of our paper-based method, but this is simply a matter of preference – the general process remains the same.)

TABLE 10.1

Asset prioritization table

ASSETS	STAKEHOLDERS' PRIORITY		
	Focus: protection. What is most important to protect from stakeholders' point of view?		**Focus: attacks.** What is most interesting/valuable for an attacker?
Description	System user	System owner	Attacker
<asset 1>	<C-? I-? A-?>	<C-? I-? A-?>	<C-? I-? A-?>
<asset 2>	<C-? I-? A-?>	<C-? I-? A-?>	<C-? I-? A-?>
...	...	...	...

In a small group where the participants know one another well, it may be more cost-effective with respect to time to let each participant express their ideas orally (one at a time), and assign one participant (or facilitator) to write the ideas on a whiteboard as they are presented. In this case, duplicates are avoided, and it may be easier (quicker) to get new ideas regarding assets based on the assets that are being presented.

10.2.2.3 Step 2: Assets from Existing Documentation

Once the brainstorming session is finished, it is a good idea to examine any available functional requirements or functional descriptions of the system to determine whether any important assets have been overlooked. In some cases, this may inspire a second round of brainstorming.

10.2.2.4 Step 3: Categorization and Prioritization

Once a list of assets is available, the assets must be categorized and prioritized with respect to security. This should be performed by the same group that participated in the asset identification, possibly augmented by management participation. (Management should possibly *not* be invited to the brainstorming session itself, since one of the pitfalls identified by Wilson [7] is inviting "anyone [...] who is feared by the other members.")

We recommend assigning priorities from three stakeholders' perspectives:

- The System User

- The System Owner

- The Attacker

As can be seen from Table 10.1, the different stakeholders' priority of the assets is described with three letters:

C: Confidentiality

I: Integrity

A: Availability

These letters can then get assigned a value indicating the importance of confidentiality, integrity, or availability for this asset. We maintain that the traditional CIA triad is enough for this purpose – additional properties like nonrepudiation will represent special cases that should be treated separately.

1. Decide which categories to use to represent priorities. In most cases it will be adequate with three (qualitative) levels:

   H:High

   M:Medium

   L:Low

2. For each asset, make a judgment regarding the different stakeholders' priority of the confidentiality, integrity, and availability of this asset. If, for example, confidentiality is not an issue for an asset, it is not necessary to include this property and assign a level to it (the same goes for integrity and availability).

3. After values for the importance of CIA have been assigned for all stakeholders of all assets, calculate a ranking sum for each security property of each asset by adding 3 for every High, 2 for every Medium, and 1 for every Low. For instance, if the asset "Web Server" is listed as A-M, C-M I-H A-H, C-M I-H A-H, the web server gets a Confidentiality rank of $0+2+2 = 4$, and so forth.

4. Create three ordered lists with the assets prioritized according to Confidentiality, Integrity, and Availability.

5. If many assets have been identified, consider pruning the assets with consistently low priorities.

If management has not been involved up to this point, they should be given the opportunity to comment on the asset tables. For small systems where a limited number of assets are identified, the final prioritization may be performed manually.

The success of the method is very much dependent on the individuals participating in asset identification, as the types of assets identified will depend on their competence and main focus. Since the method is based on brainstorming, which is not a structured method, it may be beneficial to "tune" the process by utilizing checklists or predefined questions in the brainstorming activity.

Using functional requirements as a starting point comes with the risk of not covering abstract assets such as the company's reputation, the safety of employees, and availability and connectivity of resources. We believe, however, that for this lightweight approach, most of the assets of this type can be indirectly covered by looking at the different actors' value of the assets. When stating the value of an asset from the owner's point of view, reputation should be part of the evaluation.

In the SODA approach, information on an asset is limited to values representing the importance of the confidentiality, integrity, and availability of this asset. This is done to keep the method lightweight, and it is what is needed for prioritization. The reasons behind these values and the criteria used are however lost. This may reduce the possibility to reuse the results later, and makes it harder to compare results of different sessions since the criteria may be different.

An advantage of including the attacker's perspective is that this indirectly covers some assets that are otherwise easily overlooked (or difficult to relate to). A specific example of this is the "Reputation" asset: In the past, it may not have mattered to a company if someone uses their file servers without permission for storing data, as long as they behave themselves and do not create problems for legitimate users. (The example is somewhat construed, since I cannot imagine a system administrator ever "not caring" about illegitimate users on her system; however, this effectively would have been the reaction of the local police force should the incident have been reported: "Sooo... nothing was *actually* stolen ... ?") However, this changes dramatically if the company risks being exposed in the tabloids as a haven for file-sharers and other deviants – suddenly protecting the "File Server" asset becomes much more important, implicitly because of the "Reputation" asset.

10.2.3 Example

We will now illustrate how the asset identification technique works through development of an imaginary example, which we will call the LyeFish tool.

The idea behind LyeFish is to provide a tool for amateur chefs who want to experiment with the effect of lye solutions on ichthyoids. It is basically a publicly accessible web resource with open content, but where users can log on to receive more personalized service. LyeFish shall assist the users in selecting techniques by offering a set of questions about the user's preferences and create a profile based on the answers.

LyeFish shall be an independent and self-contained application – meaning that it does not depend on any other systems to be useful to its users. However, LyeFish may contain recommendations of other products that are useful for applying the individual techniques.

The security objectives (ref. Section 10.3.2) that provided the context for the asset identification were

• Integrity of the application

• Hosting organization's IT regulations and security policy

A brainstorming session as described in Section 10.2.2.2 produces the assets

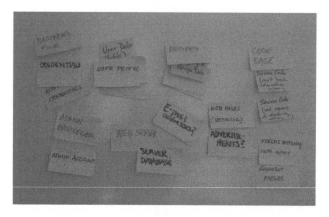

FIGURE 10.2
Result of the brainstorming session.

TABLE 10.2
Asset prioritization table

ASSETS	STAKEHOLDERS' PRIORITY		
	Focus: protection. What is most important to protect from stakeholders' point of view?		**Focus: attacks.** What is most interesting/valuable for an attacker?
Description	**System user**	**System owner**	**Attacker**
Code base		C-L I-M A-M	C-M I-H A-L
Data	I-M A-M	I-H A-M	I-L A-L
Profile	C-M I-H A-L	C-M I-H A-L	C-L I-L A-L
Credentials	C-H I-H A-L	C-H I-H A-L	C-H I-H A-L
Admin. account	C-H I-H A-H	C-H I-H A-M	C-M I-H A-L
Web Server	A-M	C-M I-H A-H	C-M I-H A-H

depicted in Figure 10.2. We then proceed immediately to classification and prioritization.

The identified assets and the results of the prioritization process are shown in Table 10.2. Reputation was also identified as an asset, but it proved difficult to fit this into the mold, and it was therefore left out of the table. However, it was agreed that damage to the integrity of the application would also damage our reputation (as system owner).

As can be seen, availability of the personalized service is not considered of high importance, since the open content will still be available. Confidentiality is also not considered relevant for the open content.

Finally, prioritization is performed for each of the security categories Confidentiality, Integrity, and Availability by assigning each H the value 3, each

TABLE 10.3

Calculated asset ranking

Confidentiality		Integrity		Availability	
ASSETS	**Sum**	**ASSETS**	**Sum**	**ASSETS**	**Sum**
Admin. account	9	Admin. account	9	Web Server	8
Credentials	9	Credentials	9	Admin. account	6
Profile	5	Profile	7	Data	5
Web Server	4	Web Server	6	Code base	5
Code base	3	Data	6	Credentials	3
Data	0	Code base	5	Profile	3

M the value 2 and each L the value 1. A missing category for any stakeholder counts as 0. This produces three prioritized lists of assets; on per security category (C,I,A), where the highest sum gives the highest priority.

In this example, note that the confidentiality and integrity of the administration account is considered equivalent (in terms of ranking score) to the other credentials in the system, but as security professionals, we intuitively give the administration account slightly higher priority. Note also that when it comes to availability, the web server itself comes out on top – this is reasonable, since the web application may deliver considerable benefit even if the administrator is unable to access it.

10.3 Security Requirements

10.3.1 Description

Information security requirements are important in all software engineering projects, not only to ensure the correct level of security in the end product, but also to avoid implementing security solutions that turn out to be a bad fit. Since security thus is important also for "ordinary" software development projects, we need mechanisms for security requirements elicitation that will be palatable to "regular" software developers and suitable for use in all software development. These mechanisms must be both easy to understand, and easy to use! Although formal methods undoubtedly have their merits, their use is precluded in this context.

Before we dive into the description of the security requirements process, we will briefly describe some artefacts which are typically used and/or created when eliciting security requirements.

Misuse cases (see Figure 10.3) [8] extend the regular use case diagrams with negative use cases (misuse cases) that specify behavior not wanted in the

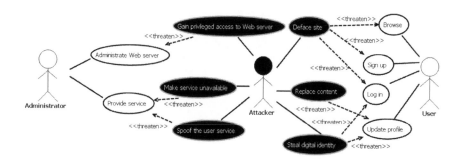

FIGURE 10.3
Misuse case diagram for a publicly available web application.

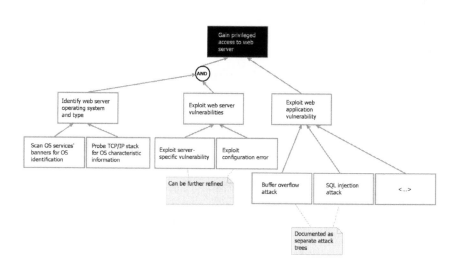

FIGURE 10.4
Attack tree detailing an attack on a web server.

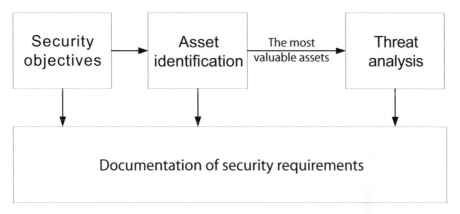

FIGURE 10.5
Core requirements phase.

system. Use cases can *mitigate* misuse cases, meaning that a use case can be a countermeasure against a misuse case – thereby reducing the chances that the misuse case succeeds. Misuse cases can *threaten* a use case, meaning that the use case is exploited or hindered by a misuse case.

Abuser stories were first introduced by Peeters [9] as an agile counterpart to misuse cases. An abuser story is a brief and informal description of how an attacker may abuse the system at hand. Abuser stories are not yet widely used, but there are some experience reports [10, 11] that show that variants have been used successfully in agile software development projects.

Attack trees Attack trees [12] represent attacks/threats against a system in a tree structure, with the goal as the root node and different ways of achieving that goal as leaf nodes (see Figure 10.4). The diagrams can be used both in the requirements and design phases. Trees can be represented graphically or can be written in outline form. It is also possible to add information on, for example, cost of attack.

The main focus of the SODA security requirements phase is identification of security objectives, assets, and threats; this results in the steps for identification of security requirements that are illustrated in Figure 10.5.

10.3.2 Security Objectives

The aim of this step is to identify the paramount security requirements; that is, the requirements that are most important to customers, and the requirements that must be met to comply with relevant legislation, policies, and standards. This is necessary to set boundaries and constraints in order to prioritize security efforts and make necessary trade-offs later. Identifying security objectives consists of two main activities: Identification of the customer's need for security, and identification of relevant legislation, policies, standards, and best practices that apply to the system/module. An important part of the process of eliciting security requirements is customer meetings that focus on security issues, and we will provide concrete tips when it comes to how to prepare for such a meeting. We also give examples of where to look for requirements from legislation, policies, and standards.

10.3.3 Asset Identification

When an asset identification has been performed (see Section 10.2), you should know which assets are most important to protect, and hopefully also which *properties* of these assets are the most important. This information should be used to prioritize requirements and to identify where to focus the effort when it comes to identifying more detailed requirements.

10.3.4 Threat Analysis and Modeling

Swiderski and Snyder [13] list the following purposes of threat modeling:

- Understand the threat profile of a system.

- Provide recommendations/solutions for secure design and implementation.

- Discover potential vulnerabilities.

- Provide feedback for the application security life cycle.

The aim here is to identify the main threats to the system/module. We recommend basing the identification of threats on the most important assets identified and the STRIDE categories (spoofing, tampering, repudiation, information disclosure, denial of service, elevation of privilege). The most important threats identified should then be further elaborated using attack trees. The attack trees identified at this stage should not be making too many assumptions regarding design decisions that have not been made yet. The attack trees will therefore be less detailed than what may be needed in later stages of the development process and may therefore need to be further elaborated in later phases. While it is certainly possible to draw threat trees and misuse case diagrams by hand, special-purpose tools such as SeaMonster (see Section 10.4.4.3) often makes the job easier, particularly when the diagrams need to be updated.

If the developers have access to a vulnerability repository (see Section 10.5.4), it should be consulted here as part of the threat analysis, and also revisited in the design phase.

10.3.5 Documentation of Security Requirements

The main aims of this activity are to make the security requirements visible, show which security requirements are high priority, and arrange for traceability and follow-up of requirements. We suggest to describe all requirements in one place, either in a separate document or as part of a general requirements document, to be able to keep an overview of all requirements.

A good security requirement is similar to pornography, in that it is difficult to define, but we recognize it when we see it.[3] We recommend to describe requirements that are focused on *what* should be achieved – not *how*. Also, negative requirements ("It should no be possible to ...") should be avoided, since they are not testable. The following is an example of a good security requirement: "Only hashed passwords shall be stored in the user database." We also recommend to give each requirement an identifier, specify the source of the requirement, and state the requirement's priority. For each requirement, it should also be possible to add information on how the requirement is followed up during development (requirement tracking).

10.3.6 Variants Based on Specific Software Methodologies

The core requirements phase is, as much as possible, not tied to a specific software development methodology. This results in the recommendations being usable for a broad group of software developers, but some developers could benefit from using other techniques that fit better with their current methodology. The technique where this is most obvious is *threat analysis*, where we currently recommend attack trees. If the developer has already used UML use cases to describe functional requirements, misuse cases will probably be a better choice than attack trees. For developers using, for example, eXtreme Programming, abuser stories will probably be the better choice. Agile developers will also probably prefer a less rigid way of documenting security requirements – for instance, by using the abuser stories for this directly. For developers using these methodologies, the recommended techniques will therefore differ from what is specified as the *core requirements phase*.

10.3.7 LyeFish Example Continued

Having identified the most important assets in Section 10.2.3, we now look into possible threats to these assets before going on to specify security requirements. We do this with the help of misuse case diagrams and attack trees.

We start out by sketching some high-level properties of the LyeFish tool, normally as a coarse use case diagram (remember to avoid making assumptions about design decisions that have not yet been made!). We then perform

---

[3]Paraphrased from Potter Stewart's concurrence in the *Jacobellis vs Ohio* case (1964).

another brainstorming exercise, this time focusing on how an attacker might abuse our system, and extend our use case diagram into a misuse case diagram.

In Figure 10.3, we have presented (parts of) a misuse case diagram for the LyeFish tool. From the diagram, we can see that the actor "Administrator" administrates the web server, and generally "provides service" through the LyeFish application. The actor "User" is someone who uses the LyeFish application via a web browser, and among the many things a User might do, the diagram illustrates browsing recipes, signing up for a personalized account, logging into an existing account, and updating personal profile. On the right side of the diagram can be seen an "Attacker" actor who might be interested in various nefarious activities, including gaining privileged access to the web server, making the service unavailable, and stealing the identity of innocent Users.

In Section 10.2.3, we determined that the administrator account and the web server itself were the most important assets, and we could argue that getting privileged access to the web server is a good step on the way to compromising the administrator account. We have therefore detailed an attack tree for this in Figure 10.4. Note that neither the misuse case diagram nor the attack tree presented can be considered complete, but are illustrations which can be used as starting points. Note also that it pays to keep misuse case diagrams and attack trees small; if they get too large, it is easy to get lost. Partly for this reason, we have decomposed the attack "Exploit Web Application Vulnerability" into specific attacks "Buffer Overflow," "SQL injection," and so forth, and each of these are documented (elsewhere) as a separate attack tree.

On the left side of Figure 10.4, there are two branches connected by an "and" symbol, meaning both branches must be accomplished in order to have a successful attack. If we can minimize the possibility of exploiting a web server vulnerability, that goes a long way, and thus a reasonable requirement would be: "A regime for timely application of security updates shall be implemented for the web server." If possible, it would be even better if "timely" could be quantified better.

Since the right branch covers web application vulnerabilities, this calls for a bunch of "classical" software security requirements, for example, "All input from web users must be validated on the server side."

10.4 Secure Software Design

The SODA approach is also applicable to software development projects where architecture and design are central.

10.4.1 Security Architecture

The architecture describes a system on an abstract level, while leaving the implementation details unspecified. The traditional security architecture deals with system level security mechanisms and issues, such as security perimeters, cryptography, access control and authorization. Having this separation can be unfortunate, as the security should be embedded into the overall software and system architecture, creating a secure architecture. This should be done by showing how the architecture, and its components satisfy both the functional and nonfunctional security requirements. The software architecture should include countermeasures to compensate for vulnerabilities or inadequate assurances in individual components or cross-component interfaces, for example, by isolating components.

The architecture and the more detailed design will usually consist of a set of (hierarchical) diagrams and documents that describe the structure and behavior of software and its environment. There are several notations and techniques for creating these, with the UML flavors as the industry leader. Several security specific extensions with tool support for UML have been created.

Creating a secure architecture and design is the overall activity in this life cycle phase and relies on and iterates with requirements specification and implementation. The next sections describe techniques that are to be used as a part of architecture and design.

10.4.2 Security Design Guidelines

Security design guidelines should be considered as a broad category of theoretical information that comes in handy when creating secure applications. These typically span from less formal best practices, principles, and rules-of-thumb to different kinds of policies, rules, regulations, and standards. Howard and Lipner [14] say that secure design best practices focus on "good security hygiene" within the application. However, the challenge is to know what good hygiene is before you start doing the "dirty work." Forcing too much theoretical information about ways to incorporate security is not very efficient. To be aligned with the SODA assumptions, we have chosen to focus on two specific kinds of guidelines and best practices; namely, security design principles and security patterns.

10.4.2.1 Security Design Principles

Security design principles are a specific type of guidelines and practices. They are proven rules for improving the security posture of an application, and in order to be useful, the principles must be applied to specific problems. This is the great advantage with them since they can be identified during the requirements phase doing threat modeling. There exist a large number of such principles, and even though just reading through them once in a while will improve security consciousness, the real value is added when they are directly used to identify weaknesses and argue for architecture and implementation decisions.

The security design principles in Table 10.4 are built on the idea of simplicity and restriction [15].

10.4.2.2   Security Patterns

A security pattern is a well-understood solution to a recurring security problem, and encourages effective reuse for building in robustness. Software design patterns have become widely accepted after the Gang of Four published their very influential book [16] on this topic in the middle of the nineties, and there exists a vast number of patterns for software development; see, for example, Hillside[4] for an extensive online library. However, while some security patterns take the form of traditional design patterns, not all of them are design patterns.

Security patterns are usually divided into different types and categories, typically:

- Structural, behavioral, and creational security patterns encompass design patterns, such as those used by the Gang of Four. They include diagrams on relationships between entities and descriptions of interaction and object creation.

- Available System patterns is a subtype of structural patterns and they facilitate construction of systems which provide predictable uninterrupted access to the services and resources they offer to users.

- Protected System patterns is another sub-type of structural patterns that facilitate construction of systems which protect valuable resources against unauthorized use, disclosure, or modification.

- Antipatterns are ways of not doing things based on things that have failed in the past or invalid assumptions. An antipattern should also include a solution, for example, reference to a working pattern.

- A mini-pattern is a shorter, less formal discussion of security expertise in terms of just a problem and its solution. Programming language-specific patterns are also known as idioms.

- Procedural patterns are patterns that can be used to improve the process for development of security-critical software. They often impact the organization or management of a development project and are therefore not security design patterns.

Table 10.5 gives examples of common security patterns, their type, and a short description.

---

[4]http://www.hillside.net/patterns/.

TABLE 10.4

Examples of design principles

Principle	Definition
Principle of Least Privilege	The *principle of least privilege* states that a subject should be given only those privileges that it needs.
Principle of Fail-Safe Defaults	The *principle of fail-safe defaults* states that, unless a subject is given explicit access to an object, it should be denied access.
Principle of Economy of Mechanism	The *principle of economy of mechanism* states that security mechanisms should be as simple as possible.
Principle of Complete Mediation	The *principle of complete mediation* requires that all accesses to objects be checked to ensure that they are allowed.
Principle of Open Design	The *principle of open design* states that the security of a mechanism should not depend on the secrecy of its design.
Principle of Separation of Privilege	The *principle of separation of privilege* states that a system should not grant permission based on a single condition.
Principle of Least Common Mechanism	The *principle of least common mechanism* states that mechanisms used to access resources should not be shared.
Principle of Psychological Acceptability	The *principle of psychological acceptability* states that security mechanisms should not make the resource more difficult to access than if the security mechanisms were not present.

TABLE 10.5

Security pattern examples

Name(s)	Type	Abstract
Single access point, Login window, One way in, Guard door, Validation screen	Protected system pattern	*Providing a security module and a way to log into the system. Set up only one way to get into the system, and if necessary, create a mechanism for deciding which subapplications to launch.*
Account lockout, Disabled password	Behavioral pattern	*Account lockout protects customer accounts from automated password-guessing attacks, by implementing a limit on incorrect password attempts before further attempts are disallowed.*
Standby, disaster recovery backup site	Available system pattern	*Structures a system so that the service provided by one component can be resumed from a different component.*
Maginot line	Antipattern	*To use a security solution that worked in the past, but is now outdated.*
Share responsibility for security, Nonseparation of duty	Procedural pattern	*This pattern makes all developers building an application responsible for the security of the system.*

10.4.3   Threat Modeling and Security Design Review

Threat modeling is an iterative process, continuously revisited throughout the software life cycle. We introduced threat modeling/analysis in Section 10.3 as an important part of the requirements phase, and we return to it here.

At this point, we know more about the system we are building than at the beginning of the requirements phase, and we use this knowledge to refine our threat models, identifying what functionality and which assets an attacker can take advantage of. The software design should be evaluated from an attacker's point of view. This process will result in a threat model document that can be used by developers to identify which threats are present, and which steps should be taken to mitigate the associated risks.

An architecture and design review helps you validate the security-related design features of your application before you start the development phase. This allows you to identify and fix potential vulnerabilities before they can be exploited and before the fix requires a substantial re-engineering effort.

Security design review is a technique that can be used to discover vulnerabilities that have been overlooked earlier in the design phase of the project. Dowd et al. [17] suggest identifying the trust boundaries in the design, and identifies six main elements to review for each boundary; authentication, authorization, accountability, confidentiality, integrity, and availability. These ideas are similar to using checklists during the security design review. Table 10.6 shows a simplified version of a checklist focused on Web application security [18].

10.4.4   Putting It into Practice – More LyeFish

We now assume that we have performed asset identification and security requirements elicitation for the LyeFish tool and proceed with secure design.

10.4.4.1   Applying Security Design Principles

Memorizing all design principles you come over is of little use. Principles are a type of knowledge that can only be fully understood through experience. In order to gain such knowledge, we recommend the following approach:

- Start with a few principles at a time. Do not try to comprehend them all at a time. You can, for instance, try to pick out the three most important ones for your current project.

- Try to understand the reason for the principles. It can slow down development if you apply principles just for the sake of it, without understanding why. It can even lead to whole layers of your application that serve no real purpose. Once you understand the reasoning behind the principles, it becomes much easier to choose how and where to apply principles.

- The most important thing is that you try - the rest comes with experience.

TABLE 10.6

Checklist for security review

Input validation	All entry points and trust boundaries are identified by the design.
	Input validation is applied whenever input is received from outside the current trust boundary.
	The design addresses potential SQL injection issues.
	The design addresses potential cross-site scripting issues.
	The design does not rely on client-side validation.
Authentication	The design partitions the Web site into public and restricted areas.
	The design identifies the mechanisms to protect the credentials over the wire (SSL, encryption, and so on).
	Account management policies are taken into consideration by the design.
	The design ensures that minimum error information is returned in the event of authentication failure.
	The identity that is used to authenticate with the database is identified by the design.
	The design adopts a policy of using least-privileged accounts.
Authorization	The role design offers sufficient separation of privileges (the design considers authorization granularity).
	The design identifies code access security requirements. Privileged resources and privileged operations are identified.
	All identities that are used by the application are identified and the resources accessed by each identity are known.
Sensitive data	The design identifies the methodology to store secrets securely.
	The design identifies protection mechanisms for sensitive data that are sent over the network.
	Secrets are not stored unless necessary.
Cryptography	The methodology to secure the encryption keys is identified.
	Platform-level cryptography is used and it has no custom implementations.
	The design identifies the key recycle policy for the application.
Exceptions	The design outlines a standardized approach to structured exception handling across the application.
	The design identifies generic error messages that are returned to the client.
Auditing & logging	The design identifies the level of auditing and logging necessary for the application and identifies the key parameters to be logged and audited.
	The design identifies the storage, security, and analysis of the application log files.

You will make mistakes, but your programs should still be superior to the ones you developed before you started thinking about principles.

For the LyeFish tool, it is natural to start with the first principle in Table 10.4, the *principle of least privilege*. We can apply this on several levels, from making sure that the LyeFish application can run as an unprivileged process, to not granting various users access to more information than they need. The *principle of fail-safe defaults* also applies, since we don't want, for example, errors in the application to give external users direct access to privileged web server content.

10.4.4.2 Making Use of Security Design Patterns

As already mentioned, several existing security patterns can be found in books, articles, and on the Web. The challenge is making these more readily available for developers, primarily from within their development tools. We see procedural patterns more as guidelines, so for the design phase, we focus on design patterns for security. We have defined two practical uses for these, namely:

- Using a security design pattern as a starting template when creating new design documents.

- Applying security design patterns to your existing design documents.

Several of the leading CASE tools support the instantiation of diagrams based on design patterns. This functionality allows you to create design stubs, making it faster and easier to make use of the proven good solutions. However, security design patterns are unfortunately not found among the default patterns for most of these tools.

For LyeFish, the two first entries in Table 10.5 are clearly relevant, in that we have the opportunity of logging into the system, and want to be able to lock out accounts that are targeted for brute-force password guessing.

10.4.4.3 Make Use of Tools for Threat Modeling

SODA threat modeling during design builds on the initial threat modeling from the requirements phase. This basis should be used as input to help select specific security design principles, guidelines, and patterns. To aid this process, it can be useful to use tools such as SeaMonster,[5] which supports several types of threat modeling notations/artifacts. Throughout the design phase, threat modeling should be continuously revisited in order to identify new threats or to mitigate the existing ones. This way, the threat model itself can be used as documentation for the security activities and countermeasures applied during development life cycle. SeaMonster facilitates reuse and sharing of threat models, and supports linking to external resources for information about threats and attacks (compliant with SODA assumption 2).

---

[5]http://sourceforge.net/projects/seamonster/.

10.4.4.4 Performing Security Review

A security and architecture review should be performed by someone else other than the designers of the target system. If there exists a dedicated security team or someone with that assigned role, that would be the obvious choice, but such a review could also be performed by regular designers and developers (with a little help). The main point is to have unbiased eyes look at and question the design artifacts that have been produced so far.

If there exist architecture and design documentation, a good start would be to go through these and see which security features are reflected here. The next step would then be to go through your findings with one of the designers to check whether your results match the intention.

If the documentation is sparse, outdated or nonexisting, you should separately interview two or more designers about the security features and compare their responses. If there are many differences, put the designers together and go through the mismatches one by one.

You do not have to wait until the end of a phase to perform a security review. A review can (and should) be performed early and several times in order to detect design flaws as soon as possible.

10.5 Testing for Software Security

Traditional software testing is mainly an exercise in Quality Assurance; "Does the application meet all the [functional] requirements [...] ?"[19]. If the requirements say "To get B, input A," the tester will input A, and if B is output, the test is categorized a success. Software security testing is more about testing things that *shouldn't* happen, however, and since the variations of possible incorrect input are practically infinite, you can never test all permutations. The security testing phase of SODA thus focuses on penetration testing of applications or parts of applications before deployment.

10.5.1 Background

According to McGraw [6], it is necessary to involve two different security testing approaches: Functional and adversarial (see Table 10.7). As mentioned, functional security testing of security mechanisms is not a controversial strategy, but many run-of-the-mill applications will have very few (or none) of these mechanisms, rendering the *functional* security testing a relatively manageable task. Since our focus is security testing of these "average" applications, we are thus primarily interested in the *adversarial* approach. Techniques for software security testing are thoroughly covered by several books [19, 20, 21, 22]. Several tools have also been developed to help testers, for example, tools that

TABLE 10.7

Approaches to security testing [6]

No.	Approach	Why
1	Functional security testing	To determine whether security mechanisms such as access control and cryptography settings are implemented and configured according to the requirements.
2	Adversarial security testing	To determine whether the software contains vulnerabilities by simulating an attacker's approach – based on risk-based security testing.

focus on error handling, monitoring of environmental interaction, and tools for intercepting and modifying traffic between server and client.

Software security testing tools and techniques have to constantly evolve to be able to detect vulnerabilities that can be utilized in new types of attacks. But there are even bigger challenges:

- Security testing is often neglected in development projects.

- When penetration testing is performed, software development organizations tend to treat the results as complete bug reports – when each item on the list has been crossed off, the system is considered "secure."

- There is seldom any feedback loop to facilitate learning from the penetration testing process to the development organization, which leads to the same type of software vulnerabilities being introduced in later versions of the software.

Security testing is a typical last-minute activity, and the resources available are scarce. Risk-based testing, which means focusing the testing effort on critical functions, therefore becomes important. Some types of security vulnerabilities are more serious and/or more common than others, and statistics and rankings like OWASP Top 10 and the SANS Top 25 can be used to focus testing. Some applications, or parts of applications, can also be more likely to cause problems:

- The highest risk is experienced by web facing systems, large code size applications, and new applications.

- Complexity may be an indicator for future security problems, using tools to measure cyclomatic complexity.

- Error handling routines should be an important focus in testing since many security failures occur in stressed environments. This is often neglected during testing because it is difficult to simulate such conditions.

In general, we advocate threat modeling as a basis for risk-based testing, focusing on application entry points. Traditional black box testing usually takes an outside-in approach where the testers do not have previous knowledge of the software to be tested. Risk-based security testing, on the other hand, implies that the test process is driven by some form of risk-related input, where the risk evaluation is based on previous knowledge of the software. The risk management process gives an indication of where an attack on the newly developed software is most likely to succeed, thus testing can be focused on the most vulnerable code.

Evidence for the benefits of risk-based testing is provided by Potter and McGraw [23]. In a case study, they did both functional and risk-based testing on smart card technologies. Via the functional tests, they found that security mechanisms often were satisfactorily implemented with respect to the defined requirements. However, when the units that previously passed the functional test were exposed to a risk-based security testing approach, all of them failed! These findings emphasize the importance of structuring and prioritizing the security tests based on risk. However, Potter and McGraw also express that risk-based security testing relies on expertise and experience, something that may be a problem in small organizations since most regular developers are not primarily interested in security issues.

10.5.2   The Software Security Testing Cycle

The software security testing cycle differs from the traditional testing cycle primarily in one aspect: Security testing does not have a clear-cut fulfillment criterion, and is therefore open-ended – we can go on testing for weeks, and still not be done. Furthermore, there are frequently limited resources available for testing, and this results in a need to prioritize testing efforts. Thus, the first step in the cycle as illustrated in Figure 10.6 is to define the focus and scope of the impending security testing activity.

We acknowledge the importance of taking a risk-based approach to security in all software development phases, including software security testing. This is even more important in a lightweight approach, where there is no way all security aspects can be fully addressed and tested. Our focus is on giving concrete guidelines as to how risk-based security testing can be achieved in ordinary software engineering projects. In Section 10.5.3, we describe how concrete results from security techniques applied in the requirements, design, and implementation phases can be used as a basis for deciding what to focus on in testing activities. These risk management activities that are tied to the specific application developed should be used together with known risk factors such as complex components, web-facing components, error handling, and so forth. In addition, it should be taken into account where we typically have failed in the past.

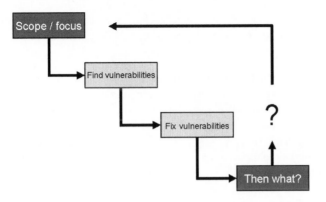

FIGURE 10.6
Software security testing cycle.

As described in Section 10.5.1, various testing techniques and tools are readily available, and we therefore do not focus on the second and third step of the testing cycle, that is, finding and fixing bugs. Instead, we choose to stress that it does not stop here! For each vulnerability that is discovered by testing, special care should be taken to verify whether similar vulnerabilities exist in other parts of the same application, and root causes identified. The results from security testing must be used to improve software development, in all phases, to make sure the same mistakes are not repeated in the next version or the next project.

10.5.3   Risk-Based Security Testing

Artifacts relevant to a risk-based testing will be produced in all phases of a software development life cycle. The lightweight techniques assisting the requirements and design phases are no different, and they output artifacts such as lists of assets, security requirements, and threat models.

The following sections describe how each of these help targeting the security testing to maximize the return on the testing effort.

Assets and Security Requirements Getting access to (or otherwise attacking) the important assets identified as part of the security requirements elicitation should be a main focus in testing activities. The security requirements will mainly give input to the functional security testing activities, but will also point to high-risk areas that should be considered for adversarial security testing.

Exploiting Threat Models in Security Testing Threat models will provide the most important input to the adversarial security testing, and help to visualize important parts of an application's attack surface; thus pointing to the areas in the code that are most exposed to attacks. Testing must be performed on these areas to ensure that the most likely attacks are not achievable.

Static Code Analysis Results Ideally, vulnerabilities reported by the static analysis tools are fixed immediately. However, it is common knowledge that these tools are not able to discover every possible vulnerability. Code segments where the returned number of vulnerabilities are above average may indicate either that the programmer has done a poor job, or that it was a particularly difficult segment to code. In both cases, one should expect more vulnerabilities to lie dormant, and thus these parts should be thoroughly tested. Also note that found vulnerabilities ("positives") sometimes are erroneously classified as "false positives."

Choice of Language Programming languages have different inherent security properties, and thus choice of language will influence the testing [24]. Since, for example, C/C++ is vulnerable to buffer overflows, this is something you would want to test for; however, this would not be relevant if, for example, Java, Pascal, or Ada were the programming language of choice.

It is also possible to take this idea a step further and look at which development environments that are used. Content Management Systems (CMS), for instance, are popular tools for web application development, and several of these have been reported to contain vulnerabilities. By searching through vulnerability databases (see below) for your development environment, a list of already known vulnerabilities can be found. These vulnerabilities should be tested to determine whether your application is at risk.

Dynamic Code Analysis and Fuzzing There are also various dynamic code analysis tools on the market, which can be useful if the development team has access to them. In particular, fuzzing tools [25] can be a golden opportunity to narrow the "infinite permutation gap" mentioned earlier (i.e., that testers can never manually test all possible combinations of input to a given program). Fuzzing is a modern variant of what used to be called "the kindergarten test"[6] in some circles. The fuzzer application will input random data via various interfaces in an automated fashion, and will record any unexpected behavior and/or failures.

10.5.4 Managing Vulnerabilities in SODA

Information about publicly known vulnerabilities are available at several sources like the Security Focus vulnerability database and the associated mailing list *Bugtraq*, the National Vulnerability Database hosted by NIST and sponsored by the Department of Homeland Security/US-CERT, and the

---

[6]Type random gibberish at the prompt, and see what happens.

Open Source Vulnerability Database. Common Vulnerabilities and Exposures (CVE), hosted by the MITRE Corporation, provides common identifiers for vulnerabilities and exposures. For a more thorough treatment of public vulnerability repositories and mailing lists, see Ardi et al.[26].

Knowledge about publicly known vulnerabilities is important and can be utilized by software development organizations. Different organizations have different characteristics, however; culture, knowledge level, or special properties of the applications being developed can be the reason why some vulnerabilities are more common than others. Knowledge of what are the most commonly introduced vulnerabilities can be used to focus testing, but also to improve the software development process.

By analyzing vulnerabilities, it is possible to understand what the organization is doing wrong, and compensate through the use of secure development techniques. To better enable learning from our own mistakes, we suggest organizing information about all vulnerabilities found in a vulnerability repository. This repository should include vulnerabilities found during security testing, but also design flaws found during design review, coding mistakes found by static analysis tools or code reviews (if applied), and vulnerabilities found after deployment. It is important to learn from vulnerabilities regardless of when they are detected.

The goals for the vulnerability repository include:

- **Improve the ability to produce secure software:** By using the vulnerability repository actively to guide the security development process in the organization, it should be possible to reduce the number of vulnerabilities in software, especially vulnerabilities that have traditionally been common or have been given focus because of the high risk involved.

- **More cost-effective handling of vulnerabilities:** Focus on common vulnerabilities should result in these vulnerabilities being avoided altogether or detected at earlier stages where the cost of fixing vulnerabilities is at a minimum.

- **Measure progress:** The vulnerability repository can be used to measure progress, that is, whether vulnerabilities are detected sooner in the development process, or whether the number of vulnerabilities are reduced. Such measurements can be a motivation factor. Note however that too much focus on reduced number of vulnerabilities can result in less effort when it comes to finding vulnerabilities, and thereby reduced quality of testing. Finding many vulnerabilities is a good thing and should also be appreciated.

Information from the vulnerability repository should be utilized at all stages of the development process in order to avoid or detect the vulnerabilities as early as possible:

- **Requirements engineering:** Vulnerabilities that are common and potentially high risk can be used as input to general software development policies that apply to all applications developed by the organization. Example: All applications shall have proper input handling. These general policies will then be used as a basis for security requirements together with customer requirements, asset identification, and threat modeling activities [2].

- **Design:** Knowledge of common vulnerabilities can guide software designers to make more secure design choices, like choosing appropriate security design guidelines, principles, and patterns. As an example: Statistical evidence that an organization's software is susceptible to SQL injection attacks may be used as an argument to include the Intercepting Validator security pattern defined by Steel et al. [27]. A design review should also profitably be based on past experiences.

- **Implementation:** The choice of language and the use of frameworks can influence which types of implementation vulnerabilities that are common. If large groups of vulnerabilities can be removed by changing, for example, programming language this should be taken into account when making such decisions. Knowledge of common vulnerabilities can also be used to focus code reviews, if the organization is performing such reviews. Finally, knowledge of common implementation errors can be used to tune the rule-sets used in the static code analysis tools.

- **Testing:** Information on where we typically have failed in the past should be used as input when prioritizing testing efforts. Concrete vulnerabilities as well as statistics on which vulnerability categories are most common should be utilized. This type of information can be obtained from the suggested vulnerability repository.

If lack of knowledge and training is identified as a cause for important groups of vulnerabilities, this information should be used to focus training initiatives. Concrete vulnerabilities should then be utilized for motivation.

To limit the work related to registering and follow-up of vulnerabilities, we suggest to strive toward only registering the information you intend to use. We suggest the following information as a minimum:

- Date – to be able to measure progress over time.

- Vulnerability category – to be able to know what areas to focus on in improvement activities.

- Where – to be able to know what portions of the code and what aspects of the application to focus on when improving security development techniques.

- In which phase and with which technique it was detected – to make sure the vulnerability is tested for in the future and to be able to see whether the same type of issue is detected earlier in future projects.

- Root cause – to use as input to improve the security development process.

- Risk – to prioritize what to focus on, but also to see if the risk posed by the vulnerabilities detected decreases over time.

- Countermeasure – to learn how this type of vulnerability can be prevented or avoided, for example, by referring to a relevant security pattern.

In addition, it may be of interest to record who introduced the vulnerability to allow each developer to learn directly from their own mistakes. However, it is important to consider possible side effects, such as employees feeling they are being publicly embarrassed and become uncomfortable with their working environment.

For the registration of vulnerabilities, we suggest to utilize predefined categories to ease aggregation and searchability of information, and use free text for details. Regarding vulnerability category, it is advantageous to use existing vulnerability taxonomies like the 7+1 kingdoms defined by Tsipenyuk et al. [28], and detail these if necessary. For describing risk it is possible to utilize the Common Vulnerability Scoring System [29]. By representing the vulnerabilities in a standard way, it will be easier to share vulnerability information in an anonymized and generalized form, so that they can be integrated in a public or federated repository.

10.5.5 Example – Testing LyeFish

We will now apply the risk-based testing approach to the LyeFish tool. In Table 10.3, we found that the administrator account and the web server itself were the highest-ranked assets of LyeFish. The administrator login interface will therefore be a logical starting point for adversarial testing.

The attack tree in Figure 10.4 next practically gives us a step-by-step procedure for attacking the web server. There are also automated tools that can be wielded against particular web servers. We can also point a fuzzer at the LyeFish web interface, and leave it running for a given time, recording all identified problems.

Finally, identified vulnerabilities (i.e., tests that compromise security) are recorded in the vulnerability repository.

10.6 Summary

With an increasing number of threats to software, security must be considered from the very beginning of every software development project. However, most software security tools and methodologies have been created with traditional security-critical projects in mind. With security becoming a concern in average projects, these methods are not always appropriate, especially for developers without the proper security background. Our motivation has been to establish a lightweight approach that should be used in every project, without consuming more resources than necessary.

This chapter has presented a lightweight approach to identifying assets, eliciting security requirements, performing secure software design, and finally security testing. Secure coding could have been a chapter of its own, but the interested reader is encouraged to explore this further in the references [5, 6]).

10.7   Further Reading and Web Sites

The SHIELDS project worked on detecting known security vulnerabilities from within design and development tools, and the project results are archived at http://shields-project.eu. This web site also contains many other useful links.

There are a number of databases of software vulnerabilities available on the internet; we consider the most important to be the *National Vulnerability Database* (http://nvd.nist.gov/), the *Open Source Vulnerability Database* (http://osvdb.org/), and the *Common Vulnerabilities and Exposures* (http://cve.mitre.org). In addition, there is a lot of useful, although less structured, information available in the *Security Focus vulnerability Database* at http://www.securityfocus.com/archive/1.

The Open Web Application Security Project maintains the "OWASP Top 10" list of the ten most critical web application security risks at http://www.owasp.org/index.php/OWASP\Top_Ten_Project. This list is augmented by the "CWE/SANS Top 25'" most dangerous software errors at http://www.sans.org/top25-software-errors/. Combined, the OWASP Top 10 and the CWE/SANS Top 25 represent a minimum baseline that all software engineering projects should be aware of in order to avoid embarrassing security errors.

Bibliography

[1] Martin Gilje Jaatun and Inger Anne Tøndel. Covering your assets in software engineering. In *The Third International Conference on Availability, Reliability and Security (ARES 2008)*, pages 1172–1179, Barcelona, Spain, 2008.

[2] Inger Anne Tøndel, Martin Gilje Jaatun, and Per Håkon Meland. Security Requirements for the Rest of Us: A Survey. *IEEE Software*, 25(1), 2008.

[3] Per Håkon Meland and Jostein Jensen. Secure software design in practice. In *Availability, Reliability and Security, (ARES 2008). Third International Conference on*, pages 1164–1171, Barcelona, Spain, March 2008.

[4] Inger Anne Tøndel, Martin Gilje Jaatun, and Jostein Jensen. Learning from software security testing. In *Software Testing Verification and Validation Workshop, 2008, ICSTW'08*, pages 286–294, April 2008.

[5] Michael Howard and David LeBlanc. *Writing Secure Code*. Microsoft Press, 2nd edition, 2003.

[6] Gary McGraw. *Software Security–Building Security In*. Addison-Wesley, 2006.

[7] Chauncey E. Wilson. Brainstorming pitfalls and best practices. *interactions*, 13(5):50–63, 2006.

[8] Guttorm Sindre and Andreas L. Opdahl. Eliciting security requirements with misuse cases. *Requirements Engineering*, 10(1):34–44, 2005.

[9] Johan Peeters. Agile Security Requirements Engineering. In *Proceedings of The 2005 Symposium on Requirements Engineering for Information Security (SREIS)*, 2005.

[10] Gustav Boström, Jaana Wäyrynen, Marine Bodén, Konstantin Beznosov, and Philippe Kruchten. Extending XP practices to support security requirements engineering. In *SESS '06: Proceedings of the 2006 International Workshop on Software Engineering for Secure Systems*, pages 11–18, New York, NY, USA, 2006. ACM Press.

[11] Vidar Kongsli. Towards agile security in web applications. In *Companion to the 21st ACM SIGPLAN Symposium on Object-oriented Programming Systems, Languages, and Applications*, OOPSLA '06, pages 805–808, New York, NY, USA, 2006. ACM.

[12] Bruce Schneier. Attack Trees–Modeling security threats. *Dr. Dobb's Journal*, July 2001.

[13] Frank Swidersky and Window Snyder. *Threat Modeling*. Microsoft Professional, 2004.

[14] Michael Howard and Steve Lipner. *The Security Development Lifecycle*. Microsoft Press, 2006.

[15] M. A. Bishop. *Computer Security: Art and Science*. Addison-Wesley Longman Publishing Co., Inc., Boston, MA, USA, 2002.

[16] E. Gamma, R. Helm, R. Johnson, and J. Vlissides. *Design Patterns: Elements of Reusable Object-Oriented Software*. Addison-Wesley Professional, 1995.

[17] M. Dowd, J. McDonald, and J. Schuh. *The Art of Software Security Assessment*. Addison-Wesley, 2007.

[18] J. D. Meier. Web application security engineering. *IEEE Security and Privacy*, 4(4):16 – 24, 2006.

[19] Chris Wysopal, Lucas Nelson, Dino Dai Zovi, and Elfriede Dustin. *The Art of Software Security Testing: Identifying Software Security Flaws*. Symantec Press, 2006.

[20] James A. Whittaker and Herbert H. Thompson. *How to Break Software Security*. Addison-Wesley, 2003.

[21] Greg Hoglund and Gary McGraw. *Exploiting Software: How to Break Code*. Addison-Wesley, 2004.

[22] Tom Gallagher, Lawrence Landauer, and Bryan Jeffries. *Hunting Security Bugs*. Microsoft Press, 2006.

[23] Bruce Potter and Gary McGraw. Software security testing. *Security & Privacy Magazine, IEEE*, 2(5):81–85, Sept.-Oct. 2004.

[24] K. M. Goertzel, T. Winograd, H. L. McKinley, L. Oh, M. Colon, T. McGibbon, E. Fedchak, and R. Vienneau. Software Security Assurance. Technical report, Information Assurance Technology Analysis Center and Data (IATAC) and Analysis Center for Software (DACS), 2007.

[25] Barton P. Miller, Lars Fredriksen, and Bryan So. An empirical study of the reliability of UNIX utilities. *Communications of the ACM*, 33(12):32–44, December 1990.

[26] Shanai Ardi, David Byers, Per Håkon Meland, Inger Anne Tøndel, and Nahid Shahmehri. How can the developer benefit from security modeling? In *The Second International Conference on Availability, Reliability and Security (ARES 2008)*, Vienna, Austria, 2007.

[27] C. Steel, R. Nagappan, and R. Lai. *Core Security Patterns: Best Practices and Strategies for J2EE(TM), Web Services, and Identity Management*. Prentice Hall, 2005.

[28] K. Tsipenyuk, B. Chess, and G. McGraw. Seven pernicious kingdoms: a taxonomy of software security errors. *Security & Privacy Magazine, IEEE*, 3(6):81–84, Nov.-Dec. 2005.

[29] Peter Mell, Karen Scarfone, and Sasha Romanosky. Common vulnerability scoring system. *Security & Privacy Magazine, IEEE*, 4(6):85–89, Nov.–Dec. 2006.

11

ICT Security Evaluation

S. J. Knapskog
NTNU, Trondheim

CONTENTS

11.1   Introduction

Security is increasingly seen as one of the basic qualities for ICT services. Without adequate security, a number of potential service users will decline the use of net-based services that they otherwise would have found to be effective and useful. Service providers must be able to convince users of the fact that information that is exchanged as a part of the service related procedures and that may be seen as sensitive, for example for economical or personal reasons, is not going astray or falling victim of any kind of abuse or misuse. However, it is not at all easy to describe and characterize ICT security in quantitative and absolute terms–the answer to this challenge may perhaps be sought with other means. It may be that adequate *assurance* that an ICT product, also often referred to as a *system*, best can be obtained by a thorough scrutiny by specialist personnel. This form of quality control will frequently be referred to as a *security evaluation* by which a technical process is performed in a security evaluation laboratory by experts. The result of the evaluation will be a technical report describing security-relevant findings. The evaluation steps will be described in detail in the current *security evaluation standards* described in the following sections, and the final verdict will be *passed* or *failed*. The laboratory itself must be organized and run by an administratively and economically independent third party and be accredited for the task by a national (governmental) overseeing body. The overseeing body will also be responsible for national certifications schemes build on the results of the security evaluations. A successful evaluation and certification framework aims at providing developers, manufacturers, vendors, and end users alike a common understanding and description of the security challenges they all are facing, and to use this framework to their advantage to describe technical and organizational measures necessary to meet the security challenges.

11.2 ISO/IEC 15408, Part 1/3 Evaluation Criteria for IT Security (CC)

11.2.1 The Development of the Standard

Evaluation of the security of ICT systems for nonmilitary applications has been performed since the beginning of the 1980s, based on the criteria published in the US standard called *The Orange Book*, with the official title: *Trusted Computer Security Evaluation Criteria (TCSEC)*. Toward the end of the decade, also Canada and a group of European countries, encompassing United Kingdom, Germany, France, and the Netherlands, had begun the development and publication of evaluation criteria intended for use in their respective national schemes for ICT security evaluation and certification. Both the Canadian and European criteria were somewhat different from the TCSEC in structure and content, since their intentions were to more strongly emphasize product evaluations than what had until then been the prevalent mode of operation used by the US evaluation scheme. As soon as product evaluation becomes the main focus area, it falls more naturally to regard security functionality and security assurance as two independent security aspects that can be specified independently, at least to a certain (some will argue fairly high) degree. The introduction of this pivotal principle opened up for a far more flexible evaluation regime, with significant potential for time-saving procedures for the actual evaluation performance, and the possibility for future procedures opening up for development of secure products by reusing previously evaluated products as building blocks when composing a more complex product or system. The Canadian and European initiatives spurred further development of the US criteria, and in the early 1990s, a document entitled *Minimum Security Functionality Requirements (MSFR)* was released to the public. This was the forerunner of a completely revised set of criteria for the US scenario, the *Federal Criteria*, which was intended to completely replace the Orange Book. The Federal Criteria incorporated the principle of independence between security functionality and assurance, and was in that respect an obvious and conscious adaptation to the development triggered by Canada and Europe toward a new evaluation paradigm.

In parallel with this transition, a development process was started in ISO, managed by the newly established Sub Committee 27 *Security Techniques (SC 27)*. The structure of the emerging standard was clearly influenced by the direction of the general trend indicated by the aforementioned national and European initiatives within the field. It had become obvious that a significant number of independent, possibly diverging, standards in the industrialized part of the world in this area could lead to a suboptimal situation both for developers, vendors, and end-users of secure ICT products and systems. It would be in the best interest of all parties involved that a worldwide internationally recognized regime for evaluation, and possibly security certification as well, could be established. This would allow the market operators to have the necessary confidence that there would be sufficient end-user demand for standardized security measures in ICT products. Standardized *evaluation criteria* and standardized *evaluation methodology* would ensure adequate scope and quality of security countermeasures, and the evaluation market would exhibit sustainable growth. At an SC 27 meeting in Stockholm in April 1990, a dedicated Working Group (WG 3) was established with the mandate to work for the future international standardization within the area. Naturally, the starting point of the work were to be the existing published national criteria. A major task was to identify the parts of these which represented the best current practice for security evaluation, both in principle, method, and technique. Once this was established, the challenge was to combine these such that a new set of internationally recognized security evaluation criteria could be agreed on by the voting members of ISO.

The nations actively involved in the development of the previous versions of evaluation criteria continued their work with the existing documents in parallel with the ISO working group. After a presentations of the newly published US Federal Criteria in Europe in 1993, a project for a joint United States, Canadian, and European Task Force named the Common Criteria Editorial Board (CCEB) was established to coordinate and further develop the parts of the existing criteria documents with the widest support, with the aim to produce one common set of documents that could be used as input documents to the international standardization process by then managed by the ISO Working Group. The project activity of CCEB has the whole time since been carefully coordinated with the standardization process in SC 27/WG 3 both time-wise and in content through a so-called Category C Liaison, signifying technical cooperation on a minute detailed level. The ISO/IEC 15408 international standard with the colloquial name *Common Criteria (CC)* was finally published in 1999, after having collected majority support from the voting members of ISO and IEC active in the subcommittee ISO/IEC JTC 1/ SC 27.[1]

11.2.2 Evaluation Model

ICT security evaluation is a technical discipline, and needs to follow the general guidelines for (today's) "best engineering practice" within the field. There are two main directions or practices for ICT security evaluations–concurrent evaluation as opposed to system evaluation after the development phase has been finalized. The distinction between an ICT product and an ICT system is defined in the criteria: a product is still sitting on the shelf of a manufacturer or a vendor. The future operative environment for the product can be assumed, but most details about the operating environment must be seen as unknown. On the other side, an ICT system is a product or set of products installed in its operating environment, and this environment is known in every necessary detail, and physical, personnel, and organizational conditions can be parameterized and taken into consideration during the evaluation.

Figure 11.1 shows the general model of an evaluation situation. The TOE (Target of Evaluation) is developed under the influence of a set of generic security requirements specified in a Protection Profile (PP) and/or specific security requirements specified in a Security Target (ST) developed in accordance with the criteria (CC). The requirements for the evaluation process itself are also found in the CC. These will in its turn be sent to an evaluation task force, together with the necessary documentation, that is, the documentation of the detailed technical procedures in the different product development phases, the product manuals for the installation and maintenance of the product in its operating environment and the user manuals for the TOE.

An evaluation report is produced as part of the evaluation task, and its finalization marks the fulfillment of the assignment. The report can be used as a basis for a subsequent certification process, but it is also naturally required by the user or owner of the TOE. In an ideal world, data could be collected in the operative phase of the lifetime of the TOE, and the security-relevant part of such data could be fed back to the different development stages of future versions of the products to improve the protection offered by the implemented security countermeasures against experienced threats present in the operational environment of the TOE. However, how to organize such closed life cycle loops in a commercial setting is still an open issue.

11.2.3 Security Requirements

The TOE incorporates security measures derived from the security objectives of the TOE. The security objectives need to be satisfied by the collection of the security requirements derived from different sources, such as

- the security policy of the organization

- identifiable threats

- laws

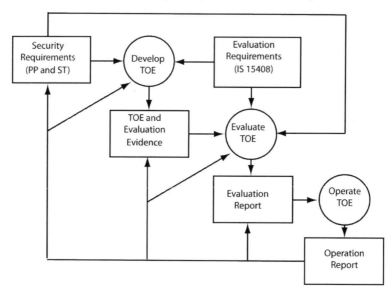

FIGURE 11.1
General model for evaluation [1].

• regulations

in addition to the knowledge and expertise found in the environment, which, in the worst case, could be used to exploit weak or missing security counter-measures or hitherto unknown vulnerabilities of the TOE. Documentation of the security objectives is done on a relative abstract level of the specification hierarchy, with increasing concretization and level of detail in the specification of security requirements, the TOE design specification and the TOE imple-mentation documentation. Some security requirements needs to be tested to be able to decide what are relevant security countermeasures in the form of security services and mechanisms. It is important to be aware of the fact that specified functional properties and behavior can be tested to the full extent, but the absence of unwanted properties or behavior never can be exhaustively tested.

11.3 Definition of Assurance

Assurance is based on Security Assurance Requirements (SAR), which are formulated in a standardized language to ensure exactness and facilitate com-parability between evaluation results. The Security Target for a TOE provides a structured description of the evaluation activities to determine correctness of the Security Assurance Requirements (SARs).

Security assurance is represented by a set of assurance components that serve as standard templates on which to base assurance requirements for TOEs. In [2] a catalog of the set of assurance components are given. In the standard the components are shown as organized into families and classes. There are seven predefined assurance packages, which are called Evaluation Assurance Levels (EALs).

If the SARs are met, there exists assurance in the correctness of the TOE, and the TOE is therefore less likely to contain vulnerabilities that can be exploited by attackers. The amount of assurance that exists in the correctness of the TOE is determined by the depth and rigor of the examinations that are to be performed according to the components required to match the SARs, and the scope of the SARs for the TOE in question.

11.4 Building Confidence in the Evaluation Process

The confidence that the security countermeasures designed and built into the TOE are as effective and appropriate as claimed by the manufacturer and/or vendor, and that they are correctly implemented must be deduced from the detailed knowledge of the product or system. The general knowledge must encompass the definition, construction, implementation, and, in the ideal case, the operation of the TOE. In a product evaluation paradigm, the information of the operating environment is normally not accessible, and the knowledge of the operating phase therefore cannot be included in the evaluator knowledge base. The evaluator can make assumptions of the future operating environment of the TOE, and base his assessment on the realism of these assumptions. The confidence that the totality of the security properties of the TOE indeed is adequate for its intended purpose must in any case be transferred from the evaluator (after the laboratory itself is satisfied that this is the case) to the end user. One of the main arguments for using a common set of evaluation criteria is that it may contribute to a common understanding of the evaluation process with its capabilities and limitations, and for the different roles the different actors are presumed to play, both in connection with the evaluation itself and the transfer of security confidence.

Evaluation techniques and methods that can be used to build assurance may encompass

- Analysis and control of processes and procedures

- Controlling whether processes and procedures are actually used

- Analysis of correspondence between product design representations at different levels of abstractions

- Analysis of TOE design documentation versus requirements contained in the system specification

- Verification of evidence

- Analysis of user documentation (users in this context may be both system operators and end users)

- Analysis of functional tests developed and the test results presented

- Independent functional testing

- Vulnerability analysis

- Penetration testing

  The list is not necessarily exhaustive.

11.5   Organizing the Requirements in the CC

The security requirements described in the CC are hierarchically ordered. The top level is called a *Class*, encompassing functional or assurance components sharing a common intent, but with different coverage for the security objectives. Security objectives are expressed by *Families*. A family is defined for those security components that aim to satisfy similar objectives, but with varying degrees of importance and thoroughness expressed by components. A *component* is a mapping of a set of security requirements, while the lowest level in this hierarchy is an *element*. An element describes atomic security requirements, that is, requirements where further subdivision would probably not lead to any meaningful evaluation result.

11.6   Assurance Elements

An assurance element belongs to one out of three *element types*:

- Elements that describe an activity performed by the developer (.D-type elements).

- Content-and-presentation of evidence, that is, what information is necessary to convince the evaluator that a claim is true, and in what form should this information be presented. (.C-type elements).

- Elements describing activity performed by the evaluator (.E-type elements). These elements shall explicitly describe whether the evaluator can confirm that the requirements for content-and-presentation of evidence has been fulfilled, and, in addition, whether the evaluator himself has performed the required assessments, tests, and analyses in addition to checking that the actions performed by the developer have been properly done and documented.

The actions performed, and the content-and-presentation of evidence presented by the developer, are influenced by the assurance requirements. These requirements form the basis for the developers demonstration of confidence that the security functionality of the TOE is adequate. In some cases, the coverage needed is expressed in the generic requirements of a *Protection profile* (PP) and the TOE-specific security requirements described in a *Security Target* (ST). An ST is preferably derived from a PP, however, it is possible to write an ST from scratch, if no suitable PP is found. The activities performed by the evaluator is guided by the evaluator's responsibilities with respect to validating the PP and ST in accordance with the Assurance Classes APE (Protection Profile Evaluation) and ASE (Security Target Evaluation). Subsequently, the verification of the claim that the TOE is correct and relates to the functional and assurance requirements expressed in the PP/ST takes place. In general, this is the basis for the confidence that the TOE is capable to meet the overall security requirements specified for it.

The generic structure of the hierarchy for assurance components is shown in Figure 11.2 The hierarchy for functional components is of similar form, but with other upper limits for the number of units at each level of the hierarchy.

An Evaluation Assurance Level (EAL) is characterized by

Scope what parts of the ICT system are security relevant and therefore must be included in the evaluation.

Depth the evaluation is performed in varying detail in design and implementation, and the appurtenant documentation for each category.

Rigor the evaluation is performed with varying emphasis on structure and formality.

A higher EAL is reached by increasing emphasis on scope, depth, and rigor during the evaluation task.

11.7 Functional Classes

The following functionality classes are defined in [4]:

- Security Audit (FAU)

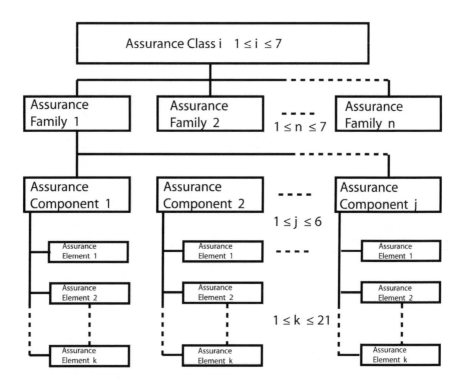

FIGURE 11.2
Generic hierarchy for the assurance components.

Assurance Class	Assurance Family	Assurance Components by Evaluation Assurance Level						
		EAL1	EAL2	EAL3	EAL4	EAL5	EAL6	EAL7
Development	ADV-ARC		1	1	1	1	1	1
	ADV-FSP	1	2	3	4	5	5	6
	ADV-IMP				1	1	2	2
	ADV-INT					2	3	3
	ADV-SPM						1	1
	ADV-TDS		1	2	3	4	5	6
Guidance Documents	AGD-OPE	1	1	1	1	1	1	1
	AGD-PRE	1	1	1	1	1	1	1
Life-cycle Support	ALC-CMC	1	2	3	4	4	5	5
	ALC-CMS	1	2	3	4	5	5	5
	ALC-DEL		1	1	1	1	1	1
	ALC-DVS			1	1	1	2	2
	ALC-FLR							
	ALC-LCD			1	1	1	1	2
	ALC-TAT				1	2	3	3
Security Target Evaluation	ASE-CCL	1	1	1	1	1	1	1
	ASE-ECD	1	1	1	1	1	1	1
	ASE-INT	1	1	1	1	1	1	1
	ASE-OBJ	1	2	2	2	2	2	2
	ASE-REQ	1	2	2	2	2	2	2
	ASE-SPD		1	1	1	1	1	1
	ASE-TSS	1	1	1	1	1	1	1
Tests	ATE-COV		1	2	2	2	3	3
	ATE-DPT			1	2	3	3	4
	ATE-FUN		1	1	1	1	2	2
	ATE-IND	1	2	2	2	2	2	3
Vulnerability Assessment	AVA-VAN	1	2	2	3	4	5	5

FIGURE 11.3

EAL 1–7 described by assurance components [3].

- Communication (FCO)

- Cryptographic Support (FCS)

- User Data Protection (FDP)

- Identification and Authentication (FIA)

- Security Management (FMT)

- Privacy (FPR)

- Protection of the TSF (FPT)

- Resource Utilization (FRU)

- TOE Access (FTA)

- Trusted Paths/Channels (FTP)

The requirement for functional security components from one or more of these classes will be expressed in a PP and/or a ST. The totality of the specified components characterize the security relevant behavior of the TOE, where relevance is given by the necessary and adequate measures to be taken to satisfy the security objectives stated for the TOE (product or system). The user will be able to detect the security behavior of the TOE by direct interaction with the TOE via its external interfaces or by observing the TOE's response to external stimuli. The set of security functionality classes is considered "open", in the sense that it can be extended by new or amended classes whenever needed, for example, triggered by new or changed future requirements.

11.8   Protection Profiles (PPs)

A Protection Profile (PP) is a generic security specification containing a set of security requirements, either taken from the CC or explicitly expressed in a separate security specification that can be assumed to be adequately addressing the security objectives of a certain type of applications. A PP describes both functional security requirements as a combined list of functional security classes, families, or components, as well as assurance requirements compliant with a given EAL. In addition to the security requirements, a PP will also contain a rationale for the security objectives, which are specified, and the corresponding security requirements, which are found necessary and adequate to satisfy these objectives.

The template ToC of a PP is shown as a list below.

- PP Introduction

- PP reference
- TOE overview

- Conformance claims

  - CC conformance claim
  - PP claim
  - Conformance rationale
  - Conformance statement

- Security problem definition

  - Threats
  - Organizational security policies
  - Assumptions

- Security objectives

  - Security objectives for the TOE
  - Security objectives for the operating environment
  - Security objectives rationale

- Extended components definition

- Security requirements

  - Security functional requirements
  - Security assurance requirements
  - Security requirements rationale

Using extended component definitions allows users to specify functional and assurance components not already defined in the CC Part 2 or Part 3 documents. This can be necessary if users (developers) conclude that the existing component sets are not quite adequate for the intended usage of the product in question, for example, if specific new threat scenarios emerge.

When specifying the security assurance requirements for an EAL, only one component from each assurance family will be chosen. The assurance components are strictly hierarchical–a component from the same family with a higher number will include all assurance elements present in a component with a lower number.

For each family used, the PP describes the actions the developer (or manufacturer) and evaluator will have to perform to establish the necessary confidence that the security measures for the TOE actually are satisfactorily established. For some of the families, the security objectives to be addressed will also be specifically mentioned in the detailed description of the EAL, if necessary, with supplementary comments given as application notes. The last part of a PP is a paragraph containing the rationale for the security objectives chosen for the TOE, what functional and assurance requirements that have been derived to obtain these, and what strength of the security mechanisms that is to be used. A PP can (should) be separately evaluated. A specific assurance class (APE) is defined for such evaluations. The objective of the APE assurance class is to demonstrate that the PP is sound and internally consistent. If the PP is based on one or more other PPs or on previously constructed component packages, it is necessary to present a reasonable rationale for the case that the PP is a correct instantiation of these PPs and packages. These properties are again necessary for the PP to be a suitable starting point for writing an ST or another PP.

11.9 Protection Profile Registries

A PP is assumed to be reusable. To disseminate the knowledge of which PPs that already are developed, and in some cases also evaluated and certified, an open register of PPs has been developed [5]. End users, organizations, companies, or special-interest groups can use this register of PPs directly if the entries therein are found to adequately address their security needs. A previously registered (potentially evaluated and certified) PP may also serve well as a starting point for the further development of new PPs, which may cover the security needs for other, possibly related, application areas with slightly different or extended security requirements. A PP that is listed in the register with status evaluated (certified) has been evaluated based on the same criteria as other ICT products or systems, that is, the CC, Part 2 and 3. In Figure 11.4, an example of a Table of Content (ToC) for a Certification Report for a PP for a Firewall is shown.

The full Certification Report is a short document (14 pages), while the Firewall PP specification is more voluminous (68 pages) following the template given in the CC, Part 1. Both documents are available at [5]. At the same location, a list of certified products is to be found [6].

11.10 Definition of a Security Target (ST)

The Security Target (ST) can be seen as an instantiation of a Protection Profile. In most evaluation schemes, this is the usual and recommended organization of the evaluation tasks. However, all TOEs are not covered by a generic security specification, that is, a PP, hence an ST may have to be con-

```
┌─────────────────────────────────────────────────┐
│                                                 │
│     Certification Report - Firewall protection profile │
│                                                 │
│                      TOC                        │
│                                          Page   │
│   1. Summary                              1     │
│   2. Information for identification        3     │
│   3. Security policies                     4     │
│   4. Assumptions and scope                 5     │
│         4.1 Assumptions                    5     │
│         4.2 Scope to counter threats       6     │
│   5. PP information                        7     │
│         5.1 Security functional requirements 7   │
│         5.2 Assurance packages             8     │
│   6. Evaluation results                    9     │
│   7. Recommendations                      10     │
│   8. Acronyms                             10     │
│   9. References                           11     │
│                                                 │
└─────────────────────────────────────────────────┘
```

FIGURE 11.4
The ToC of a certification report for a Firewall PP.

structed from scratch. In Figure 11.5, the standard structural outline of the ST is given, although alternative structures are allowed. For instance, if the security requirements rationale is particularly bulky, it could be included in an appendix of the ST instead of in the security requirements section.

The separate sections of an ST are summarily described below.

The ST introduction shall contain three narrative descriptions of the TOE on different levels of abstraction. The conformance claim part shall make it clear whether the ST claims conformance to any PPs and/or packages, and if so, to which PPs and/or packages. Then a security problem definition is mandated, showing threats, under what Organizational Security Policies (OSPs) the TOE is supposed to work, and related assumptions, which are of relevance for the holistic security assessment. A vital part of the ST is, of course, the description of the security objectives, showing how the solutions to the identified or assumed security problems are divided between security objectives for the TOE and security objectives for the operational environment of the TOE. In a given situation, an extended components definition wherein new components (that is, those not included in CC Part 2 or CC Part 3) may be deemed necessary, and, consequently, will have to be separately defined in order to be able to handle extended functional and extended assurance requirements. The ensuing description of the security requirements can be seen as a translation of the security objectives for the TOE into a standardized language. This standardized language is in the form of Security Functional Requirements (SFRs). Additionally, this section defines the Security Assur-

Security Target

- **ST Introduction**
  - ○ ST reference
  - ○ TOE Reference
  - ○ TOE overview
  - ○ TOE Description
- **Conformance claims**
  - ○ CC conformance claim
  - ○ PP claim
  - ○ Package claim
  - ○ Conformance rationale
- **Security problem definition**
  - ○ Threats
  - ○ Organizational security policies
  - ○ Assumptions
- **Security objectives**
  - ○ Security objectives for the operating environment
- **Extended components definition**
- **Security requirements**
  - ○ Security functional requirements
  - ○ Security assurance requirements
  - ○ Security requirements rationale
- **TOE summary specification**

FIGURE 11.5

The ToC of a security target (ST) for a TOE [7].

ance Requirements (SARs). The TOE summary specification, showing how the SFRs are implemented in the TOE, is rounding off the ST.

For specific TOEs claiming low assurance, there exists low-assurance STs that have reduced contents; these are described in detail in section A.12 of the CC part 1. All other security specifications aiming at an ensuing evaluation situation assume an ST with full contents.

Before and during the TOE evaluation, the ST specifies what is to be evaluated. In this role, the ST serves as a basis for an agreement between the developer and the evaluator on the exact security properties of the TOE and the exact scope of the evaluation. Technical correctness and completeness are major issues for this role. A TOE evaluation always starts with an evaluation of the ST itself in accordance with the criteria given in the assurance class ASE described in Part 3 of the CC. Evaluating an ST is required to demonstrate that the ST is sound and internally consistent. If the ST is based on one or more PPs or packages, a believable case should be made for the fact that the ST is a correct instantiation of a given PP or clearly identified well-defined packages. These properties are necessary for the ST to be suitable for use as the basis for a TOE evaluation.

After the TOE evaluation, the ST specifies what was evaluated. In this role, the ST serves as a basis for agreement between the developer or re-seller of the TOE and the potential consumer of the TOE. The ST describes the exact security properties of the TOE in an abstract manner, and the potential

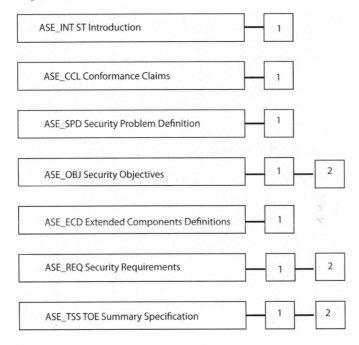

FIGURE 11.6
The structure of the assurance class ASE–security target evaluations [8].

consumer can rely on this description because the TOE has been evaluated to meet the ST. Ease of use and understandability are major issues for this role. The assurance is about the exact nature of security offered by the evaluated TOE, that is, it may be possible to offer high assurance of moderate or even low security.

11.11 Evaluation of a Security Target (ST)

The assurance class for the ST evaluation is limited in scope compared to the TOE assurance classes since there are no security features implemented in the ST itself. Nevertheless, to examine the main steps of the evaluation description as given in [8], assurance class ASE is a pivotal exercise to get a feeling for the process steps involved in the evaluation of any of the TOE assurance classes. The following text is written in close correspondence with the text in the standard itself. Figure 11.6 shows the families within the ASE class, and the hierarchy of components within the families as they appear in [8].

The evaluation of the ST is performed family by family, and for each family, there is a fixed pattern to be followed. First, the objective of the family that is to be evaluated is checked. Then the process goes on through the following additional steps:

- Checking for dependencies (between components)

- Checking developer action elements

- Checking content and presentation elements

- Checking evaluator action elements

As an example, let us look at the material in the CC Part 3 describing the evaluation of just one of the families–the ASE-REQ: Security requirements.

Objectives

- The SFRs form a clear, unambiguous, and well-defined description of the expected security behavior of the TOE. The SARs form a clear, unambiguous and canonical description of the expected activities that will be undertaken to gain assurance in the TOE.

- Evaluation of the security requirements is required to ensure that they are clear, unambiguous, and well-defined.

Component leveling The components in this family are leveled on whether they are stated as is.

ASE_REQ.1 Stated Security Requirements

Dependencies: ASE_ECD.1 Extended components definition

Developer action elements:

- ASE_REQ.1.1D The developer shall provide a statement of security requirements.

- ASE_REQ.1.2D The developer shall provide a security requirements rationale.

Content and presentation elements:

- ASE_REQ.1.1C The statement of security requirements shall describe the SFRs and the SARs.

- ASE_REQ.1.2C All subjects, objects, operations, security attributes, external entities, and other terms that are used in the SFRs and the SARs shall be defined.

- ASE_REQ.1.3C The statement of security requirements shall identify all operations on the security requirements.

- ASE_REQ.1.4C All operations shall be performed correctly.
- ASE_REQ.1.5C Each dependency of the security requirements shall either be satisfied, or the security requirements rationale shall justify the dependency not being satisfied.
- ASE_REQ.1.6C The statement of security requirements shall be internally consistent.

Evaluator action elements:

- ASE_REQ.1.1E The evaluator shall confirm that the information provided meets all requirements for content and presentation of evidence.

ASE_REQ.2 Derived Security Requirements

Dependencies: ASE_OBJ.2 Security objectives

- ASE_ECD.1 Extended components definition.

Developer action elements:

- ASE_REQ.2.1D The developer shall provide a statement of security requirements.
- ASE_REQ.2.2D The developer shall provide a security requirements rationale.

Content and presentation elements:

- ASE_REQ.2.1C The statement of security requirements shall describe the SFRs and the SARs.
- ASE_REQ.2.2C All subjects, objects, operations, security attributes, external entities, and other terms that are used in the SFRs and the SARs shall be defined.
- ASE_REQ.2.3C The statement of security requirements shall identify all operations on the security requirements.
- ASE_REQ.2.4C All operations shall be performed correctly.
- ASE_REQ.2.5C Each dependency of the security requirements shall either be satisfied, or the security requirements rationale shall justify the dependency not being satisfied.
- ASE_REQ.2.6C The security requirements rationale shall trace each SFR back to the security objectives for the TOE.
- ASE_REQ.2.7C The security requirements rationale shall demonstrate that the SFRs meet all security objectives for the TOE.
- ASE_REQ.2.8C The security requirements rationale shall explain why the SARs were chosen.

- ASE_REQ.2.9C The statement of security requirements shall be internally consistent.

Evaluator action elements:

- ASE_REQ.2.1E The evaluator shall confirm that the information provided meets all requirements for content and presentation of evidence.

This pattern is repeated for all the families in the ASE class. The result will first be used as previously mentioned as an agreement between sponsor and evaluator that the ST is a valid specification of the security objectives and requirements of the TOE, and second as a way to communicate to the end user exactly what security features present in the TOE have been evaluated to a defined assurance level (EAL).

11.12 Evaluation Schemes

In order to achieve greater comparability between evaluation results, evaluations should be performed within the framework of an authoritative evaluation scheme that sets the standards, monitors the quality of the evaluations, and administers the regulations to which the evaluation facilities and evaluators must conform. The CC does not state requirements for the regulatory framework. However, consistency between the regulatory frameworks of different evaluation authorities will be necessary to achieve the goal of mutual recognition of the results of such evaluations. A number of national evaluation schemes managed by national authorities have established mutual agreements of acceptance of each other's evaluation results and product certificates. These are the so-called CCRA countries. The list of countries presently members of the CCRA is given at the CC portal [9].

A second way of achieving greater comparability between evaluation results is to mandate (recommend) the evaluation facilities to use a common methodology to achieve these results. For the CC, this methodology is given in [10]. Use of a common evaluation methodology in general contributes to the repeatability and objectivity of the evaluation results but is not sufficient by itself. Many of the evaluation criteria require the application of expert judgment and background knowledge for which consistency is more difficult to achieve. In order to enhance the consistency of the evaluation findings even further than can be expected by using the criteria and the CEM, the final evaluation results may be submitted to a certification process. The certification process is the independent inspection of the results of the evaluation leading to the production of the final certificate or approval, which is normally publicly available. The certification process is a means of gaining greater consistency in the application of IT security criteria. Members of the CCRA arrangement also have certification authorities who will recognize one another's certificates on a mutual basis. The evaluation schemes and certification processes are the responsibility of the evaluation authorities which run such schemes and processes and are outside the scope of the CC.

11.13 Evaluation Methodology

The target audience for the Common Methodology for Information Technology Security Evaluation (CEM) [10] is primarily evaluators applying the CC and certifiers confirming evaluator actions. A secondary interest group may be evaluation sponsors, developers, PP/ST authors, and other parties interested in IT security. The CEM defines the minimum actions to be performed by an evaluator in order to conduct a CC evaluation. Presently, only evaluations up to and including EAL 5 is covered in the CEM document. The CEM is closely aligned with the CC, using the criteria and evaluation evidence defined therein. There are in fact direct relationships between the CC structure (i.e., class, family, component, and element) and the structure of the CEM. Figure 11.7 illustrates the correspondence between the CC constructs of class, family and evaluator action elements and CEM activities, subactivities, and actions. However, several CEM work units may result from the requirements noted in CC Developer Action Elements and the Content and Presentation Elements.

The CEM provides an overview of the evaluation process and defines the tasks an evaluator is expected to perform when conducting an evaluation. Each evaluation, whether of a PP, ST, or TOE follows the same process. There are four main tasks an evaluator laboratory has to perform for all evaluations: the input task, the output task, the evaluation subactivities, and the demonstration of their technical competence to the evaluation authority task. An introductory description of the input task and the output tasks, which are related to management of evaluation evidence and to report generation, is found in chapter 8 of the CEM. Each task has associated sub-tasks that apply to, and are normative for all CC evaluations (evaluation of a PP or a TOE). The full description of the tasks is given in the following chapters. The input and output tasks have no verdicts associated with them, as they do not map to any of the CC Evaluator Action Elements. They are performed in order to ensure conformance with the universal principles for an objective, an impartial evaluation process and to comply with the CEM. The demonstration of the technical competence for the evaluation authority task may be satisfied by the evaluation authority analysis of the output tasks results, or may include the demonstration by the evaluators of their understanding of the inputs for the evaluation subactivities. This task has no associated evaluator verdict, but has an evaluator authority verdict. The detailed criteria to pass this task are left to the discretion of the evaluation authority.

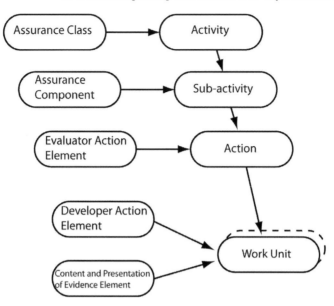

FIGURE 11.7
Relationship between CC and CEM structures [10].

Four roles are defined in the CEM: sponsor, developer, evaluator, and evaluation authority. The sponsor is responsible for requesting and supporting an evaluation. This means that the sponsor establishes the different contractual agreements with the evaluator and evaluation authority for the evaluation. Consequently, the sponsor is also responsible for ensuring that the evaluator is provided with the appropriate evaluation evidence. The developer produces the TOE and is responsible for providing the evidence required for the evaluation on behalf of the sponsor. In many cases, the sponsor and developer roles are handled by the same organization or company. The evaluator performs the evaluation tasks required in the context of an evaluation. It is the evaluator role who receives the evaluation evidence from the developer, either on behalf of the sponsor or directly from the sponsor, performs the evaluation sub-activities presents the results of the evaluation, and proposes an assessment verdict to the evaluation authority. The evaluation authority role establishes and maintains the scheme, monitors the evaluation conducted by the evaluator by assessing Observation Reports (ORs) and taking an active part in resolving questions raised in the ORs. The evaluation authority is also responsible for the approval of the Evaluation Technical Report (ETR), which contains the final evaluation verdict recommended by the evaluator. Subsequently, the evaluation authority may issue a certificate based on the final evaluation verdict contained in the ETR. To prevent any undue influence from affecting an evaluation in an improper way, a strict separation of roles is required. This implies that the roles described above are handled by different entities, except the roles of developer and sponsor which may be satisfied by a single entity, as stated previously. In some low-assurance evaluations (e.g., EAL1) the developer may even not be required to be involved in the project at all. In this case, it is the sponsor who provides the TOE for the evaluator and who will have to generate and submit the evaluation evidence to the evaluator.

11.14   Summary

In this chapter, the principles and practices used for evaluating the security of ICT products, whether they are implemented in software or hardware, whether they are small or large, simple or complex, have been described. The evaluation must be done according to the international standard ISO/IEC 15408 Evaluation Criteria for IT Security, Parts 1–3, which is equivalent in content to the Common Criteria (CC), Parts 1–3. Unlike most ISO/IEC standards, the CC is freely downloadable from the net (see below). The main features of the evaluation methods, are described in an accompanying standard, entitled *Information Technology–Security Techniques–Methodology for IT Security Evaluation*, which is in fact a technical report. The text of this standard is equivalent to the content of the technical report entitled *the Common Criteria Evaluation Methodology (CEM)*, which also can be downloaded for free at the address given below. Together, the criteria and methodology standards are the necessary background material needed to set up and operate an evaluation laboratory, which, given that an appropriately trained workforce is engaged, and is capable of performing ICT security evaluations that in turn can be used as a basis for security certification of ICT products and systems by an accredited and internationally recognized *Certification Body*. At the time of this writing, a number of industrialized nations, some of which are members of the *Common Criteria Recognition agreement (CCRA)*, already have this ICT security evaluation infrastructure in place, and operative.

11.15   Further Reading and Web Sites

Contrary to most ISO standards, both the ISO/IEC 15408, Parts 1–3 and the ISO/IEC 18045 standards are freely downloadable from the ISO ITTF Portal: http://standards.iso.org/ittf/PubliclyAvailableStandards/ Documents with equivalent text are also obtainable at the Common Criteria Portal: http://www.commoncriteriaportal.org under the headings CC,

Parts 1–3 and CEM. Presently, the CC versions are numbered 3.1R3, indicating that three minor revisions of the documents have been performed since the last major revision (from V2 to V3). The CC Portal web-site also contains further information about the CCRA and other reading material related to ICT security evaluations.

Bibliography

[1] ISO/IEC SC27/WG 3 Experts. *ISO/IEC 15408-1:1999. Information Technology – Security Techniques – Evaluation Criteria for IT Security – Part 1: General Model.* ISO, October 1999. `http://standards.iso.org/ittf/PubliclyAvailableStandards/`.

[2] ISO/IEC SC27/WG 3 Experts. *ISO/IEC 15408-2:2008. Information Technology - Security Techniques - Evaluation Criteria for IT Security - Part 2: Security Functional Requirements.* ISO, August 2008. `http://standards.iso.org/ittf/PubliclyAvailableStandards/`.

[3] ISO/IEC SC27/WG 3 Experts. *ISO/IEC 15408-1:2005. Information Technology – Security Techniques – Evaluation Criteria for IT Security - - Part 1: General Model.* ISO, October 2005. `http://standards.iso.org/ittf/PubliclyAvailableStandards/`.

[4] ISO/IEC SC27/WG 3 Experts. *ISO/IEC 15408 -2:2005. Information Technology – Security Techniques – Evaluation Criteria for IT Security – Part 2: Security Functional Requirements.* ISO, October 2005. `http://standards.iso.org/ittf/PubliclyAvailableStandards/`.

[5] The Common Criteria Managerial Board. *Common Criteria Protection Profiles.* NIST, June 2009. `http://www.commoncriteriaportal.org/pp.html`.

[6] The Common Criteria Managerial Board. *Certified Products List.* NIST, November 2009. `http://www.commoncriteriaportal.org/products.html`.

[7] ISO/IEC SC27/WG 3 Experts. *ISO/IEC TR 15446:2009. Information Technology – Security Techniques – Guide for the Production of Protection Profiles and Security Targets.* ISO, October 2009.

[8] ISO/IEC SC27/WG 3 Experts. *ISO/IEC 15408-3:2005. Information Technology – Security Techniques – Evaluation Criteria for IT Security – Part 3: Security Assurance Requirements.* ISO, October 2005. `http://standards.iso.org/ittf/PubliclyAvailableStandards/`.

[9] The Common Criteria Managerial Board. *The Common Criteria Recognition Agreement.* NIST, November 2009. `http://www.commoncriteriaportal.org/theccra.html`.

[10] ISO/IEC SC27/WG 3 Experts. *ISO/IEC 18045:2005. Information Technology – Security Techniques – Methodology for IT Security Evaluation.* ISO, October 2005. `http://standards.iso.org/ittf/PubliclyAvailableStandards/`.

12

ICT and Forensic Science

S. F. Mjølsnes and S. Y. Willassen
Department of Telematics, NTNU

CONTENTS

12.1   The Crime Scene

The pervasiveness of computers and networks both in private and public use in our society provides an abundance of digital evidence to the inquisitive questions: *Where were you? What happened there? Who did it?* Although acts of crime might primarily be directed at physical targets and people, they will leave digital tracks in one or more electronic devices "witnessing" the course of the incident.

In general, we can distinguish between one or more of the following roles of computers and networks in relation to public law and order:

- Computers and networks can be the direct target of *intentional incidents* caused by an perpetrator. This is challenge of information security, where we attempt to construct means to prevent and detect the attacks.

- As technical assistants and tools (processing, storing, and communication) used by the perpetrator for the purposes of vandalism, disruption, gain, and profit. This is the realm of malicious hacking and cyber warfare.

- As passive sources of evidence, or even acting as witnesses providing "technical testimonies about the incident or criminal act under investigation. This is the object of the forensic field called digital forensic science.

- As computational instruments for assisting in the investigative process of finding and analyzing physical and technical evidence. This is the ambition of computational forensics.

This chapter examines technical approaches to investigating crimes that involve electronic computing devices and digital communication in some way. The methods and knowledge of this investigative process and the tools needed for this particular type of evidence is often denoted *digital* forensic science, or *digital forensics* for short. Digital forensics is the investigation process concerned with evidence gathered from the digital tracks. *Computational forensics* involves supporting all technical investigations and evidence analysis with computational instrumentation, thereby replacing human opinions and assessments with objective measurements.

Cybercrime is a broad term for breaking national and international law in cyberspace. In this definition, *cyberspace* denotes the new territory of digital network infrastructures, where people can connect globally and augment their social and organizational affairs and business. Most of us seek to conquer and exploit this new land of opportunity and fortune, where few domestic and international laws apply yet. Many roamers of cyberspace behave well, like they do in the physical world, but some want to break with national laws. And then there are some that act like the Wild West robbers, outlaws, and snake oil peddlers, that could be called *cybercriminals*. The local sheriff has a hard time tracking down the offenders while remaining within his realm of authority and law.

The problem of attacks on and disruption of internetworked systems has escalated and will continue to do so. Information servers of companies and organizations are attacked for competitive or political reasons. A country's critical information infrastructures can been attacked and disrupted. Many investigations of internetworked attacks meet investigational challenges that are transnational and across jurisdictions. This is the *cyber security* problem. NATO foresee that even nations can clash in cyberspace, and seek out a strategy for *cyber defense*. Seven NATO nations have established the Cooperative Cyber Defence Centre of Excellence in Tallinn, Estonia [1].

The Council of Europe developed a Convention on Cybercrime [2] in 2001 that has been ratified by 43 nations. This convention, which entered into force on July 1, 2004, is the first international treaty on cybercrimes. The convention lists 11 categories of criminal activity that each signatory state must adopt into its national law. We describe these here in abbreviated form:

Illegal access unauthorized access to any part of a computer system.

Illegal interception unauthorized interception of nonpublic data communication.

Data interference unauthorized damaging, deletion, alteration, or suppression of computer data.

System interference unauthorized disruption of a computer system.

Misuse of devices

 a. Production, sale, procurement for use, import, distribution of

  i. A device designed primarily for the purpose of committing offenses.

 ii. A password or access code by which any part of computer system can be accessed for the purpose of committing offenses.

 b. Possession of such an item with the intent of being used for committing offences.

Computer-related forgery input, alteration, deletion, or suppression of computer data resulting in unauthentic data.

Computer-related fraud causing a loss of property to another person by computer data modification or interference with a computer system.

Offenses related to child pornography producing, offering, or making available, distributing or transmitting, procuring, or possessing child pornography through a computer system or on a computer-data storage medium;

Infringements of copyright and related rights where such acts are committed willfully, on a commercial scale and by means of a computer system.

Attempt and aiding and abetting with the above.

12.2 Forensic Science and ICT

The term *forensic* is derived from the latin *forum*, denoting the public square in roman cities. The Forum Romanum was the ruling hub of the Roman Empire, serving as the place for proclaiming public matters of a religious, political, and judicial nature in the empire. The proceedings of disputes and public trials were conducted orally, and the actual evidence supporting the claims was physically and methodically presented to a judge positioned on a tribunal. The evidence could take the form of testimony of witnesses, physical objects, or documents. Evidence means what is clearly there for all to see.

Basically, we are using the same system in our legal courts, whereby it is distinguished between the factual evidence available and the claims put forth by the disputing parties. Excellent rhetoric should not be sufficient to convince the judge. The acquisition, presentation, authentication, and judicial establishment of evidential facts relevant to the claims are fundamental to the legal procedure. Our legal procedure requires that the verdict is to be founded on evidence, either strongly beyond reasonable doubt in criminal cases, or with preponderant probability in civil cases.

Forensic science has been popularized for a long time through the detective novels of Poe, Doyle, Christie, and plenty more authors, and through theater, film, and now television productions. Currently, several television series, such as *Crime Scene Investigation*, are presenting fanciful technology for detection and measurements, assisting the detective's process of investigation. A recent article by Houck [3] claims that

> the effect that the CSI programs have had on the activities of police, who now collect more pieces of physical evidence than ever before; and in academia, where forensics programs are growing exponentially.

However, the gap between the fiction and the facts of existing methods in reality is significant. These fictional series show measurements on samples of evidence, resulting almost instantly in precise and accurate detection and interpretation by the forensic expert. In reality, forensic practioneers will more often than not find ambiguous answers.

Digital forensic investigation technology will become an integrated part of the design and management of the critical infrastructures. A correct and complete explanation of the causes and events of intentional incidents provides improved risk assessments, deeper understanding of the vulnerabilities, and enhanced requirements for the technological building blocks used for protection of ICT-based critical infrastructures. Equally important is the deterrence factor, as prevention against cybercrime that can be achieved by a successful digital investigative response that provides efficient analysis of digital forensic evidence. The enforcement instruments available to the judicial system are currently weak for networked critical infrastructures because there is a lack of efficient technological methodology that can assist digital *after-the-fact investigations* and, at the same time, abide by the democratic rights to personal privacy.

The tools and techniques of digital forensics can also be useful for

- Organization internal investigations of potential breaks of policies.

- Troubleshooting operational problems, for instance, resolving configuration problems.

- Auditing of event transactions logs.

- Recovery and reconstruction of lost data.

- Deleting data from equipment that is removed, to be outdated or reused in other settings.

12.3   Evidence

12.3.1   Judicial Evidence

Scientific theory enables us to assert true predictions about the future effects given the initial conditions as general as possible. We collect scientific evidence by constructing and repeating controlled experiments that may support or reject our hypothesis. We set up the initial conditions, we perform the experiment, and we study the effects of the process. This is *causal* modeling and reasoning.

Philosophically, the biggest forensic case ever raised for investigation is What made the universe? and How did it happen? If we take on the Big Bang Hypothesis proposed by physics, the observable and measurable background "noise" in space can be interpreted as after-the-fact evidence. We can still "hear" the reverbations of the event, according to physics. This evidence is not a static picture from the past, but live dynamic background "noise" originating from the event. It is dynamic evidence, but we cannot repeat the full process.

Some scientific fields, like geology, archaeology and paleontology, are mostly concerned with answering the "what happened?" question, and are barred from being able to initiate nor repeat the full processes that these scientist are studying. They are to a large degree limited to investigations by recovery, documentation, analysis, and interpretation of the remaining artifacts and environment, thereby attempting to infer past events. This is *evidential* modeling and reasoning.

A legal trial is normally occupied with attempting to answer questions about what happened, by trying to reconstruct which events and causes are most likely for the observed effects for the *particular case* in the dispute. The starting point and very foundation for the legal trial inference process is the process of preservation, collection, validation, identification, analysis, interpretation, documentation, and presentation of judicial evidence.

12.3.2 Digital Evidence

Fingerprints, footprints, blood are some of the classic evidence types. In the context of digital forensics, evidence can be electronic devices and storage media, files, memory dumps, e-mail and chat messages, and so on. The question of authenticity of the evidence can quickly become an issue in convincing the court. Even more so for digital evidence, where we know, a priori, that data can easily be modified. Therefore, evidence collection and presentation process must follow strict technical and legal requirements. Some jurisdictions, like in most US states, have established strict rules for admissibility of evidence in court. In other jurisdictions, like in many European countries, the court is not legally bound in deciding what will be accepted as evidence in each trial.

Although a file is found to be stored on a seized harddisk, the content of the file cannot be directly read or viewed physically by the people in court. The content must be communicated to the court via a display or a paper printout. This line of reasoning brings up the issue of whether digital evidence can be directly presented as evidence at all, or it must be viewed as documentation about the evidence presented. This emphasizes the importance of handling digital evidence according to rules that are acceptable to the court. The procedures and means for ensuring the authenticity and integrity of the collected digital evidence are the main topics of most existing textbooks on digital forensics.

The court must become convinced that the presented digital evidence is *authentic*, the main requirements being

Origin The source (device, owner, author, program, connection, etc.), the location, and the time of collection.

Integrity The method of extraction, measurement errors, chain-of-custody, and integrity verification techniques.

For instance, it may be reasonable to distinguish between data stored on a computer, whether the input is by a person or a computer program. It may also be relevant to distinguish between data from a live computer and data from a digital archive. As for integrity, there is an ongoing discussion about whether it is appropriate to shut off a device or not in order to preserve the integrity at seize time.

US Federal Rules of Evidence [4] state general rules for both for civil and criminal cases. The rules distinguish between original and duplicate or writing and recording, but make an expedient and pragmatic definition for computer printouts.

An "original" of a writing or recording is the writing or recording itself or any counterpart intended to have the same effect by a person executing or issuing it. An "original" of a photograph includes the negative or any print therefrom. If data are stored in a computer or similar device, any printout or other output readable by sight, shown to reflect the data accurately, is an "original."

The last sentence of this quote represents the current convention and practice by which digital forensic investigators act. However, this presupposition of *original* is hardly future proof as we move to a society where the workings of networked mobile computing devices become well understood by ordinary people and lawyers. New concepts and definitions suitable for cyberspace are needed.

12.3.3 Evidential Reasoning

The problem of what constitutes an *explanation* has been treated in philosophy and logic for a long time. In our context here, the general question is how can we make correct inferences from the evidence we collected. Well, Lett [5] has proposed some criteria that sound evidential reasoning in science should follow:

FALSIFIABILITY It must be possible to conceive of evidence that would prove the claim false.

LOGIC Any argument offered as evidence in support of any claim must be sound.

COMPREHENSIVENESS The evidence offered in support of any claim must be exhaustive – that is, all of the available evidence must be considered.

HONESTY The evidence offered in support of any claim must be evaluated without self-deception.

REPLICABILITY If the evidence for any claim is based upon an experimental result, or if the evidence offered in support of any claim could logically be explained as coincidental, then it is necessary for the evidence to be repeated in subsequent experiments or trials.

SUFFICIENCY The evidence offered in support of any claim must be adequate to establish the truth of that claim, with these stipulations:

1. The burden of proof for any claim rests on the claimant.
2. Extraordinary claims demand extraordinary evidence.
3. Evidence based upon authority and/or testimony is always inadequate for any claim.

The ideal forensic method should be based on objective scientific method and procedure, rather than subjective assessment and evaluation. The *Daubert test* came out of a United States Supreme Court case in 1993, and is a set of criteria by which to deem the theory of a forensic method acceptable or not. The requirements of Daubert are:

1. That the theory is testable (has it been tested?)
2. That the theory has been peer reviewed (peer reviewing usually reduces the chances of error in the theory)
3. The reliability and error rate (100% reliability and zero error are not required, but the rates must be reported)
4. The extent of general acceptance by the scientific community

The legal system places the *burden-of-proof* on the prosecuting party in a criminal case. To reach a guilty verdict, the court must find that the presented evidence proves the defendants guilt beyond reasonable doubt. Evidence in favor of the prosecution party is called *inculpatory*, whereas evidence in favor of the defendant's case is called *exculpatory*. These impressive words reflect the Latin word *culpa*, which simply means guilt or fault. The investigating party, the police, is obligated to search for, collect, and present both types of evidence.

12.3.4 Lack of Evidence

Consider the two corresponding notions *evidence of absence* and *absence of evidence*, and how these will apply to digital forensics. The notion "evidence of absence" follows this claim pattern:

If event E occurred in this computer, then evidence of E can be found by any clever detective.

And the equivalent negated implication reads

If no evidence of E can be found in this computer by all these clever detectives, then event E did not occur.

Many will probably follow this mode of reasoning if relevant to the case, though it is philosophically contentious whether one can positively prove something by the absence of evidence. It appears that the evidence of absence rests in the belief of the unfailing detectives that caused the absence of evidence.

12.4 The Digital Investigation Process

Investigations are inquiries into past events. The purpose of an investigation is to find evidence that can establish an understanding of previous events. During an investigation, evidence is examined in order to produce information about past events. Possible sources of evidence include witness statements, documents, physical evidence (i.e., fingerprints or biological evidence), and data stored on digital media. From examination of the evidence, information about the past events can be reconstructed. Event reconstruction is the final outcome of an investigation, and forms the basis for a decision. The final event reconstruction relies on interpretation of the evidence and is usually performed by a person or group of persons separate from those performing the investigation. This person or group is called the finder of fact and could be a judge, magistrate, jury, or other depending on the case and jurisdiction. During the investigation, the investigator formulates theories about possible events in order to find further sources of evidence, prepare the case for the correct jurisdiction, and present the evidence to the fact finder in an appropriate manner.

The qualifying word *forensic* means that the investigation is performed in the context of law, and that the pieces of evidence found and identified are presentable according to the procedures and understanding in the court of law. Digital forensic techniques are used for investigating crimes, break of organizational policy, audits, security incidents, and to re-create deleted or ruined data.

Investigation of digital devices and media with the purpose of finding evidence is commonly referred to as *digital investigation*. The purpose of digital investigation is to find digital evidence related to the events under investigation and present them to the fact finder. The Electronic Crime Scene Guide [6] states the investigation process should adhere to these principles:

- The process of collecting, securing, and transporting digital evidence should not change the evidence.

- Digital evidence should be examined only by those trained specifically for that purpose.

- Everything done during the seizure, transportation, and storage of digital evidence should be fully documented, preserved, and available for review.

The process of investigating can be divided into different phases, in general guided by the standard steps of collecting evidence at a crime scene:

1. Preservation of the digital crime scene. Locate and isolate the electronic devices from the area, preserve any volatile data, turn off the devices, make an exact copy of the data stored on the system to another medium. Seize the selected objects into custody while making sure the integrity and authenticity are carefully preserved.

2. Examine the objects and identify the obvious pieces of digital evidence, such as nondeleted documents, e-mail, and so forth. Obtain an understanding of how the system has been used.

3. Document the findings.

4. Search for and examine the digital evidence. Thorough analysis of the contents of the digital image(s). Search for evidence, including keyword search and signature search. Examination of files that may contain new types of evidence. Perform time-line analysis.

5. Digital crime scene event reconstruction. Identify how the evidence got there and what its existence means. Which events took place on the digital crime scene and who did it? Determine alternative explanations, if any, and test which explanation is most likely.

6. Presentation of a digital crime scene theory. Present the results of the digital investigation to the investigation team and if necessary, interpret them.

7. At a later stage, present the evidence and results in an acceptable form to the fact finder.

First, digital media must be found and identified. Then, data on it must be preserved in order to secure the evidential integrity. Secure the media in the order of volatility (processor registers, memory, processes, harddisk, optical disc read-only). This usually involves copying the data on the medium to another medium in such a way that no data are changed on the original. The data on the copy can then be analyzed for contents relevant as evidence.

Because of the large amounts of data stored on modern data storage devices, the search is usually performed by a combination of manual and automatic search. There exist a number of helpful techniques employed by investigators in this phase, such as keyword search, hashing, and signature search. When any relevant data have been found, it must be documented, usually in the form of a report. Finally, the data are presented to the fact finder, either as printouts, or by having the investigator appear before the fact finder to report the findings. Although final event reconstruction is up to the fact finder, one should bear in mind that the fact finder usually has little expertise in digital computing. The investigator is often asked to present his theory on how the presented evidence can be used to reconstruct the investigated events.

The documentation or *chain of custody* of the objects selected in the investigation is very important in this process in order to provide high-quality evidence acceptable to the court. The practice of this process for storage media and PCs, hard disks, PDAs, mobile phones, printers, copy and fax machines, and other electronic devices has been developed and adopted by police and others, and is regular practice and state-of-the-art now.

Reverse engineering techniques have been proven to be very helpful in problems of reconstructing data for evidence. Recently, there has been an increasing interest in more sophisticated methodologies for forensic analysis, including crime scene reconstructions, time-line analysis, and studies of evidence dynamics. It has been argued that the current approaches to digital forensics are severely limited due to the lack of coherent frameworks and approaches for digital forensics and new procedures to understand and model competing requirements.

Crime scene reconstruction is a fairly new development in forensic science, but it has gained popularity in the digital forensics research community. The purpose of the method is to determine the most probable hypothesis or sequence of events by applying scientific methods to interpret the events that surround the commission of a crime. The basic approach is to state the problem, form a hypothesis, collect data, test the hypotheses, follow up on the most promising hypothesis, and finally draw conclusions supported by admissible evidence. The analysis may involve the use logical reasoning and statistical analysis, as well as domain knowledge about people, criminology, and so forth. The conclusions of a crime scene reconstruction are usually given with a level of certainty associated with the different hypotheses, indicating the level of evidentiary value.

A formalization of the digital investigation process has been proposed by Carrier [7], by introducing the concept of an object history. For a computer, the history includes the complete set of configurations, states, and events that has occurred during the lifetime of the computer. A state is the sum of all variables that may occur in the computer, whereas an event is any action that may change the state. In this model, the digital investigation process is defined as formulating hypotheses about the history of the computer, and testing them against known values such as known user-input, data from other evidence sources, and the final state of the system. With this model, the assumptions on which the event reconstruction is based are more explicit. This makes it possible for the investigator and the fact finder to assess the assumptions and decide whether they are justified.

12.5   Digital Evidence Extraction

12.5.1   Sources of Digital Evidence

Digital storage media and devices are obvious sources of digital evidence. Most of the personal electronic devices contain persistent storage such as hard disks, compact disks, and flash memory. Magnetic hard disks are standard components of all computers, but can be found in printers, scanners, and copy machines too.

Long-term storage for archival or backup purposes include optical disc storage media (CD, DVD), magnetic storage by tape or disc, non-volatile semiconductor memory (flash, EEPROM, ROM).

Electronic handheld devices contain digital storage, including mobile phones, PDA, music players, GPS location devices, gaming devices, arm-wrist watches, photo and video cameras, and soon more wearable computers that will be part of our clothes, earplugs, goggles, and headgear.

Our houses contain electronic music instruments, digital radios and televisions, telephones, cameras and recorders, remote controllers, door locks and process control equipment, professional tools, and more.

There are all kinds of electronic devices with storage and communication services used in vehicles and transportation. And last, but certainly not the least, most of the entities and devices in telecommunication networks, local area networks, wireless and mobile access networks. The list seems endless!

12.5.2   Extraction

Image copy

Imaging of storage makes a complete copy of the hard drive (or other medium) containing possible evidence. Imaging is done with software that copies all data and computes hashes of the content. The image copy also contains deleted data. The hard drive is connected through some special hardware device that electrically will allow only reading operation, disabling possible modification to the source.

File copy

The selected files may contain deleted data, but deleted data stored elsewhere on the medium are not retained. There is a risk of disturbing data content and metadata since data on the medium are changed during a file copy.

The choice between the two methods depends on practical circumstances and of course the competence of the investigator.

12.6   Digital Evidence Analysis Techniques

The digital forensic investigator has a wide range of techniques at hand to find crucial evidence in a case. Here, we will only briefly list the most common techniques and how they work.

Keyword search By defining one or several keywords, the investigator can search through large amounts of data and quickly find information items relevant to the case. Entire hard drives seized during an investigation can typically be searched for many keywords in a matter of hours. Findings can be made inside existing files, or within remnants of previously deleted files that still exist on the medium.

Signature search In a similar way, the investigator can also search the entire medium for *file signatures*, sequences of data occurring at the beginning and end of known file formats. Such a search allows the investigators to extract all files of a certain type from large data volumes, for example, extracting all JPG images from unallocated area of the hard disk. This technique is also called *file carving*.

Hashing Hashing implies calculating a file hash for seized files using a hash algorithm such as MD5 or SHA1. This allows for simple correlation between sets of known files. The technique is typically used to identify known contraband, such as sexual abuse images. Another possible application of this techique is to exclude files known to belong to software installations. This helps the investigator to focus on the files specific to the seized computer, which saves effort.

Timestamp analysis File systems contains a plethora of timestamps. A commonly used investigation technique is to list all files on the seized medium and sort them according to the various timestamps on the file system in question. This allows the investigator to focus on a specific period in time, and determine which activity took place on the computer during that time period.

A common challenge in Digital Forensics is to handle large amounts of data. It is not uncommon in Digital Forensics to have several hard drives seized in the same case. This can amount to several terabytes worth of data, which means that data analysis quickly becomes a significant challenge. It is not an option in practice to do manual analysis of all data in real cases. This means that an investigator must have an idea of what (s)he is looking for in order to select investigation techniques and formulate keywords. This idea must come from a thorough understanding of the case under investigation, including all relevant details. Only with such knowledge can the investigator be able to define search criteria and recognize pieces of information that are relevant to the case under investigation when found on the digital medium.

12.7 Anti-Forensics

The ongoing research within tools and methods for digital investigation has resulted in efforts in creating tools and methods for *anti-forensics*. Anti-forensics can be defined as the research into and development of tools to compromise the availability or usefulness of evidence to the forensic process [8]. This can be accomplished in a variety of ways ranging from overwriting deleted data to prevent them from being discovered, through usage of strong encryption to obscure contraband data, to the introduction of directory loops in a file system that causes forensic software to deadlock or crash when reading file system structures [9].

The parallel development activity of forensic and anti-forensic tools can be thought of as an arms race in which development in one field stimulates development in the other. The development of techniques for forensic recovery of deleted information, for example, led to the development of the tool EvidenceEliminator [10]. This tool removes temporary files and overwrites areas of hard drives containing deleted material with random data. This has in turn led to the development of forensic techniques, by which specific patterns resulting from the use of EvidenceEliminator are recognized, so the investigator can prove that this particular anti-forensic tool was run [11]. Conversely, the existence of steganography techniques for hiding files within other files has inspired the development of forensic techniques that can examine files for superfluous data as well as tools for determining signatures produced by specific steganography tools [12]. A common argument is derived from extrapolating the existence of forensic – anti-forensic technique pairs into a belief that any forensic technique will ultimately be matched by an anti-forensic technique that will block evidence extraction performed with the forensic technique. This argument can be called the Arms Race Argument. Supporters of this argument believe that any new developments within forensics can only be effective for a limited amount of time, since ultimately an anti-forensic technique will prevent the forensic technique from being used effectively.

There are several fallacies with the arms race argument that reduces its weight as an argument against the effectiveness of forensic techniques. First, in most cases, it is possible to determine whether an anti-forensics program has been used. It is difficult to make an anti-forensics program in such a way that it does not leave any pattern specific to that program. For example, an anti-TimeStampLogic program would very likely produce specific patterns between timestamps it would have to change in order to perform its function. These patterns could be detected by an anti-anti-TimeStampLogic program. Although hypotheses about the original clock could then no longer be tested properly, the evidence of usage of anti-TimeStampLogic would create an impression that there was something to hide. This reduces the desirability of using an anti-forensics tool significantly.

The most serious fallacy in the arms race argument is however the underlying assumption that anti-forensic techniques will always be available and will be used by everyone possessing potential sources of digital evidence. Consider the adversaries in a digital investigation, the Investigator and the Perpetrator. The Investigator usually possesses knowledge of digital investigation and tools that can comb a digital medium for evidence, including tools for digital imaging and data recovery. The Perpetrator is, on the other hand, likely to be an average computer user, and does not know how to protect himself from the scrutiny of a digital investigation or where he can obtain the necessary tools.

The Investigator also has time on his side. Once a digital medium has been imaged by a forensically sound method, the Investigator has plenty of time to investigate its contents. The Perpetrator, on the other hand, never knows when the Investigator will turn up to seize his data, if ever. He therefore has to be prepared at all times and run the anti-forensic tools again and again after every action that would leave incriminating evidence. There is no room for mistakes by the Perpetrator. If he makes a small mistake in his anti-forensic procedures, the evidence may be there waiting to be discovered by the Investigator. The Investigator, on the other hand, can make a lot of mistakes, as long as he does not mess up the original data. He can always start from a fresh image at a later time, should he feel that there is more to find or that current results rely on misinterpretations.

All in all, the Investigator has a tremendous advantage over the Perpetrator in digital investigations. This is true in other types of investigations too, hence the saying, "There is no such thing as a perfect crime". And we must remember the fact that the widespread knowledge and use of glove "technology and tools" has not eliminated the utility of fingerprints in crime investigations.

12.8   Further Reading and Web Sites

Many textbooks and expert training books exist on computer forensics, computer evidence, cybercrime, digital investigations, and forensics, for example [13, 14, 15].

The forensic examiner needs both methods and tools. The Electronic Crime Scene Guide [6] is available online from U.S. National Institute of Justice. NIST Computer Forensic Tool Testing resources are available at `www.cftt.nist.gov`.

The *International Journal of Digital Evidence* (IJDE) contains theory, research, policy, and practice articles in the rapidly changing field of digital evidence, see `www.ijde.org`. The *Journal of Digital Forensics, Security and Law* (JDFSL) can be found at `www.jdfsl.org`.

You can read about the topic of computational forensics in a recent article in *IEEE Spectrum Magazine* [16].

Bibliography

[1] NATO. Cooperative cyber defence centre of excellence.

[2] CoE. Convention on cybercrime. Council of Europe (ETS No. 185), November 2001.

[3] M. M. Houck. CSI: reality. *Scientific American Magazine*, 295(1):84–89, 2006.

[4] Federal Evidence Review. Federal rules of evidence. U.S. Department of Justice, 2011.

[5] J. Lett. A field guide to critical thinking. *Skeptical Inquirer*, 14:153–160, 1990.

[6] NIJ. *Electronic Crime Scene Investigation: A Guide for First Responders*. National Institute of Justice, 2nd edition, 2008.

[7] B. Carrier. A hypothesis-based approach to digital forensic investigations. Technical report, Center for Education and Research in Information Assurance and Security , Purdue University Tech Report, 2006-06.

[8] R. Harris. Arriving at an anti-forensics consensus: Examining how to define and control the anti-forensics problem. *Digital Investigation*, 3:44–49, 2006.

[9] T. Newsham, C. Palmer, A. Stamos, and J. Burns. Breaking forensics software: Weaknesses in critical evidence collection. In *Proceedings of the 2007 Black Hat Conference*, 2007.

[10] Robin Hood Software Ltd,. Evidence eliminator. http://www.evidence-eliminator.com, 2011.

[11] Radsoft.net. The evidence eliminator documents. http://www.radsoft.net/resources/software/reviews/ee/05.html, 2011.

[12] J. Fridrich, M. Goljan, D. Soukal, and T. Holotyak. Forensic Steganalysis: Determining the Stego Key in Spatial Domain Steganography. In *Proceedings of SPIE Electronic Imaging*, volume 5681, pages 631–642. San Jose, CA, 2005.

[13] Eoghan Casey. *Digital Evidence and Computer Crime: Forensic Science, Computers and the Internet*. Academic Press, 2nd edition edition, February 2004. ISBN 0-121-63104-4.

[14] Brian Carrier. *File System Forensic Analysis*. Addison-Wesley Professional, March 2005. ISBN-13: 978-0-321-26817-4.

[15] M. A. Caloyannides. *Privacy Protection and Computer Forensics*. Artech House, 2 edition, 2004.

[16] Sargur N. Srihari. Beyond C.S.I.: The rise of computational forensics. *IEEE Spectrum Magazine*, 47(12):36–41, December 2010.

13

Risk Assessment

S. Haugen

Department of Production and Quality Engineering, NTNU

CONTENTS

This chapter discusses risk, risk analysis, and risk evaluation in general terms but seen in relation to information security. The purpose is to give an overview over the process in general terms and provide guidance on how this is performed, irrespective of the methods chosen.

13.1   Risk Assessment in the Risk Management Process

Risk assessment fills a central role in the information security risk management process, and it may be useful to start by illustrating how this process fits into the overall risk management process. The description is adapted from ISO/IEC 27005 [1]. This is a different way of illustrating part of the risk governance process described in Figure 13.1, mainly covering Risk Appraisal (largely corresponding to Risk Analysis in Figure 13.1), Tolerability and Acceptability Judgment (Risk Evaluation), and part of Risk Management (Risk Treatment).

A similar process is also described in the generic risk management standard ISO 31000 [2].

In this chapter, the focus is on the *core* of this process, that is, the Risk assessment process. In addition to this, definition of context and risk treatment also influences the process. These steps in the process are therefore also de-

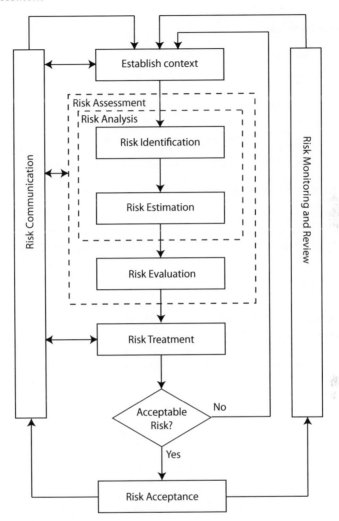

FIGURE 13.1
Overview of risk management process.

scribed briefly, in particular those aspects of the process which are important
for the risk assessment.

The terminology in this field is not always well defined and used in the same
way in different applications. This can often cause considerable confusion.
Initially, some of the key terms used in this context are therefore discussed.

13.2 Terminology

13.2.1 Risk

The term *risk* is used widely in very different contexts and frequently with different meanings. In the context of risk management in general, a common definition can be as follows [3]:

Risk is the combination of the probability of an event and its consequences

In information security, this definition is often extended [4]:

Risk is the potential that a given threat will exploit vulnerabilities of an asset or a group of assets and thereby cause harm to the organization. It is measured in terms of a combination of the probability of an event and its consequence.

A similar definition is also applied in [1].

In this context, risk is related to something that can *cause harm to the organization*, that is, it has to do with something that may have a negative outcome. It may be noted that, in other contexts, risk may also be positive. This is particularly relevant when considering financial risk, where risk usually is associated with potential for both positive and negative outcomes. Further, the definition states that risk is associated with *threats, vulnerabilities*, and also *events*, all of which will be discussed further below.

The second part of the definition states how risk is measured, that is, as *a combination of the probability of an event and its consequence*. This is often expressed as risk being equal to probability times consequence. In effect, this means that risk is an expression of the statistically expected consequence of an event. Frequency is often used instead of probability to express risk.

There are also alternative definitions of risk. Aven [5] defines risk as *Uncertainty about and severity of the consequences of an activity*. In this definition, the probability has been replaced with uncertainty, while *severity of the consequences* in reality corresponds to *conseuqences* in the earlier definition. Further, risk is tied to activity rather than events.

The consequences of an event can be of a varying nature and varying extent. In traditional risk analysis, the most common types of consequences are related to loss of life or injuries, environmental consequences, or economical consequences. For information security, these may be relevant, but there are also other types of consequences that may be more relevant to consider:

- Loss of functionality of systems

- Loss of integrity of data

- Delay in production/deliveries

- Damage to reputation

- Breach of contractual, legal, or regulatory requirements

- Release of confidential information

For some of these consequence types, it is relevant to consider not only the severity or the extent but also the duration. This applies to, for example, loss of functionality, where the time until the function is restored may have a significant effect on the *harm to the organization* (see also discussion of *vulnerability* below). This may be less relevant for other consequence types, for example, *Release of confidential information* where the effect is irreversible.

13.2.2 Vulnerability

In the definition of risk, the term *vulnerability* is also introduced, in the context of *vulnerabilities of assets*.

ISO/IEC 13335:1-2004 [4] defines vulnerability as *a weakness of an asset or a group of assets that can be exploited by one or more threats*. If this is compared with the definition of risk, this essentially only says that vulnerability is *a weakness*. More enlightening is the definition in TR19791:2006:

flaw, weakness or property of the design or implementation of an information system (including its security controls) or its environment that could be intentionally or unintentionally exploited to adversely affect an organizations assets or operations.

This widens the definition to include *flaw, weakness, or property* related to the design, implementation, or also the environment of a system. If exploited, the vulnerabilities may *adversely affect an organizations assets or operations*. It is noted that, when vulnerability is defined in this way, it is a property of the system that is being analyzed and is independent of the hazards that the system is exposed to.

13.2.3 Hazards, Threats, Sources, and Events

In the definition of risk, *threat* is introduced in the definition and so is also *event*. The term *threat* is defined as follows (ISO/IEC 13335-1:2004 [4]):

Threat is a potential cause of an incident that may result in harm to a system or an organization.

Two other terms that are also frequently used in the same context is *hazard* and *source*. *Threat* and *hazard* are two terms that are often interchanged, and *source* is sometimes also used, converying more or less the same meaning (see also definition of *risk analysis* below).

Event is not defined in ISO/IEC 13335-1:2004 [4], but incident is defined and the term event is included in the definition:

Information security incident is any unexpected event that might cause a compromise of business activities or information security.

In some cases, *event* or *incident* and *hazard* are interchanged.

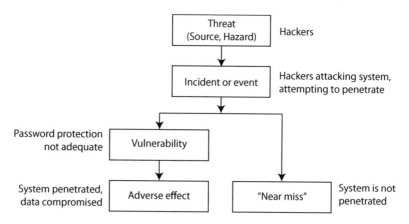

FIGURE 13.2
Illustration of key terms.

Figure 13.2 illustrates the different terms and how they can be linked together. The starting point is the *Threat*, in this case illustrated by computer hackers who may want to try to penetrate the system and damage or steal data. The threat can be seen as a situation or condition that exists continually, but as long as no attempts are made at attacking the system, this will not cause any negative effects on the system. The *incident* is a situation where the threat is realized, in this case when the hacker actually attacks the system and attempts to penetrate it. In this case, two possible outcomes of this are shown:

- If there is a vulnerability (a weakness) in the system, adverse effects may be experienced.

- Alternatively, there are no vulnerabilities that can be exploited and the controls in place successfully prevent the hacker from penetrating the system. No adverse effects are experienced.

13.2.4 Risk Analysis, Risk Evaluation, and Risk Assessment

The terms risk analysis, risk evaluation, and risk assessment also needs to be defined. The three terms and how they are related are visualized in Figure 13.2. All of these definitions are based on ISO/IEC 27001 [6] (which in turn is based on ISO/IEC Guide 73:2002 [3]):

Risk analysis is systematic use of information to identify sources and to estimate the risk.

In this definition, the term *source* is used. This could cover both *threat* and *incident* as defined above. However, in practical terms, *sources* should be the relevant incidents that may occur. *Estimate risk* will usually mean that the probability of the unwanted events and the consequences are described or estimated qualitatively or quantitatively and combined. Since threats can be present constantly (as a condition), it is not always meaningful to talk about the probability of the presence of a threat, as opposed to the probability of something happening (an event). It can be assumed that hackers exist so it does not really make sense to ask what the probability of hackers existing is, but we may ask what the probability that hackers will attack a system is.

In simple terms, risk analysis is about answering three questions:

- What can happen (what are the events/incidents)?

- What are the causes (and the probability) of this happening (what are the hazards or threats)?

- What are the consequences if this happens?

The next term that we need to define is *risk evaluation*:

Risk evaluation is the process of comparing the estimated risk against given risk criteria to determine the significance of the risk.

This definition is straightforward in the sense that it states that the results from the risk analysis (*the estimated risk*) should be compared with given criteria. *Significance of risk* can mean a lot of different things, but in this case, it is about determining whether this is a risk that we decide to accept or whether it needs to be treated (reduced). This is further discussed below.

Finally, Risk assessment is the *overall process of risk analysis and evaluation*. In other words, this is just the combined process of risk analysis and risk evaluation.

13.3 Main Elements of the Risk Assessment Process

In Figure 13.3, the risk assessment process has been expanded to illustrate the individual steps in more detail. In the following, each of the steps in the figure are briefly discussed. At the end of this section, a brief discussion on alternatives to using risk assessment in the risk management process is included.

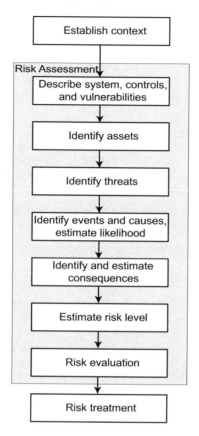

FIGURE 13.3
The overall process of risk analysis and evaluation.

13.3.1   Establish Context

Some of the key elements in establishing the context of the analysis are

- Establishing decision criteria for the risk evaluation, including risk acceptance criteria

- Defining the objective and scope of the assessment

- Defining limitations and boundaries of the assessment

- Establishing an organization/project team for performing the risk assessment

Decision criteria for use in the risk evaluation may be established as part of the overall risk management process or specifically for a particular risk assessment. Risk acceptance criteria may be formulated in different ways and the risk level that is accepted may vary significantly, depending on the context. A high risk may be accepted relating to economical risk while this is not acceptable in relation to breach of regulatory or legal requirements. Acceptable risk levels will not be discussed, but highlight some aspects of how decision criteria are formulated and how this in turn may influence the risk analysis.

Risk acceptance criteria True risk acceptance criteria will be formulated such that they take into account both the likelihood and the consequences of events, in line with the definition of risk. However, they may be expressed in different ways. One possible formulation is to express the criteria in terms of maximum acceptable risk level per event or per threat that is identified. This means that the risk analysis needs to express the risk per event and ensures that no individual events or threats represent a high risk. Another formulation is to state a maximum accepted total risk level, that is, the sum of the risk for all threats and events identified. This means that individual threats/events may have a high risk as long as there are few relevant events that are identified.

Maximum consequence criteria This is an alternative decision criterion which focuses on what the worst-case consequences are and sets criteria to limit these. This is not a risk criterion as such since the likelihood is not taken into account, but may be used in some cases because the consequences may be so catastrophic that all feasible measures which are within practical and economical limits will be put in place to avoid them.

A combination of different criteria may also be applied. The formulation of the criteria will influence how the results from the risk analysis need to be presented, and this places constraints on how the analysis is performed.

Definition of the scope and the objective of the assessment will also have to be done at this stage. It is useful to define what decisions the assessment should support. An example could be whether this is an assessment aimed at defining what controls are required in a new system (design) as opposed to assessing whether an existing system has an acceptable risk level (audit).

One aspect of defining the scope is also to define the types of consequences that are considered. In some cases, for example, it may be that only economical consequences are considered.

Setting the boundaries for what should be included in the analysis and defining the limitations of the analysis will also be part of the definition of the context. The whole organization may be the object of the assessment, parts of the organization, individual systems or specific functions performed by a system. In all cases, relationships with functions, systems, or organizations outside the boundaries of the analysis object needs to be considered.

The type of threats considered in the assessment may also be limited. Sometimes, assessments are limited only to deliberate threats, like criminal acts, sabotage, and industrial espionage. Other limitations may also be imposed due to resource limitations or for other reasons.

The decision criteria, the scope, and the limitations in the assessment will also have a major influence on the choice of method for performing the analysis.

Establishing the context also includes determining who will be involved in the preparation of the risk assessment. Usually, the risk assessment is performed by a risk assessment team. Key requirements to the team are that they need to know the organization or system that is being analyzed, how it functions, and what controls are in place. Knowledge of the method that is used for risk assessment is also necessary.

Risk assessments will have to be repeated periodically, to assess the effect of changes. The system may be modified, the way it is used may change, and there may also be changes both in threats and vulnerabilities. All of these changes may contribute to changing the risk and revisiting this with regular intervals is necessary, to ensure that appropriate controls are in place. Criteria for when the risk assessment should be updated may be established as part of the risk management system. Criteria may be established covering updates:

- after a given time period

- when functionality is implemented or removed

- when hardware or software is modified

- when procedures and processes are changed

13.3.2   Describe System, Controls, and Vulnerabilities

The next step is to describe the system that is to be analyzed. The description can be structured in different ways, but it is often useful to break down the system into smaller units or functions, as a basis for a structured analysis of each individual part. Following the breakdown, it may be helpful to define the following:

- What are the most critical functions undertaken by each system part?

- What is the input required for the system part to function?

- What is the output from the system part?

To understand how risk is managed in the system, existing controls to reduce risk and vulnerabilities should be identified. Controls are safeguards and risk-reducing measures that are put in place to avoid adverse effects to the organization. The controls may be grouped according to how they contribute to managing risks, that is,

- Controls that contribute to reduce or remove threats. This could be, for example, locating the hardware in a location where water flooding is not possible.

- Controls that contribute to reduce the likelihood of events or incidents, for example, use of nonflammable material.

- Controls that contribute to reduce the consequences of events, for example, backup power generators that can be used in power outages.

Controls may be of a very different nature and some examples include:

- Organizational measures such as training of personnel to increase awareness of the importance of password security.

- Procedures that describe how systems are operated in a safe manner.

- Physical measures, such as locked doors preventing unauthorized access to assets.

- Software measures, such as limitation on attempted logins before being shut out of the system.

Suggestions for relevant controls can among others be found in ISO/IEC 27001/2:2005 [6]. Identification of existing controls can be done by:

- Review of documents containing information about controls already in place and controls planned put in place.

- Interviewing those responsible for information security, to determine which controls are in place and are effective.

- On-site review and inspection of physical controls.

As part of the system description, the vulnerabilities of the system should also be identified. Vulnerabilities will not automatically have a negative effect on the system being considered, but represent a weakness or a deficiency that may be exploited by a threat. Similarly, a threat will not have a negative effect on the system unless it is combined with a vulnerability.

Vulnerabilities may be identified in a number of possible areas:

- Organization

- Processes and procedures

- Personnel

- Physical environment

- Information system configuration

- Hardware, software, or communications equipment

- Dependence on external parties

Vulnerabilities may be related to different situations:

- There may be vulnerabilities associated with the way the system is supposed to operate, because there are deficits in the design, operational procedures, or other. Examples could be that there are security holes in software which were not identified during development.

- There may be vulnerabilities because the system is being used in a different way than originally intended, without proper analysis of the effect of the changes. An example could be that a system is extended, without changing backup routines such that backup is not performed on the new parts of the system.

- There may be vulnerabilities associated with controls that are not working as intended, either not working at all, or with inadequate reliability. This could be, for example, a UPS (Uninterruptible Power Supply, also known as battery backup) that has a low reliability and may fail when needed.

13.3.3 Identify Assets

The identification of assets involves identification of what is of value to the organization and that the organization therefore wants to protect. Assets may include physical objects, information, and intangible assets. Examples of assets are

- physical assets, for example, computer hardware, communication facilities, buildings, other equipment

- software

- information/data

- people

- goodwill, reputation and so forth

The degree of detail to which assets are identified and described will depend on the scope of the analysis. A high-level identification of assets may be the first step, followed by more detailed identification for the most critical assets.

Assets should be described with respect to:

- Maximum consequence (criticality) for the organization if the asset is lost or loses it functionality. This should be expressed both in terms of economical value and other intangible costs.

- Possibility of replacing the asset and how long it will take to replace it.

13.3.4 Identify Threats

The identification of threats can be done in different ways and using different sources, but the key is to identify all significant threats that the organization needs to protect its assets against. There are different approaches that can be taken to do this, and combinations of approaches are often used. Possible approaches to the threat identification are

- Use of checklists/standardized lists of threats. An example of such a list can be found in ISO/IEC 27005:2008 [1]. Industry bodies and government bodies may also provide checklists.

- Review of past experience, for example records of incidents and events that have happened earlier.

- Review of earlier risk assessments for similar systems

- Workshops/workgroup meetings where a creative process to identify threats is applied

- Inspections

- Predictions about future developments

When using checklists and past experience it is important to consider whether this is relevant for the system or organization being studied. The threat identification should focus on what may happen in the future, not just consider past experience.

Threats are often split in random (accidental) and deliberate threats. The strategy for reducing risk has to be very different to protect against deliberate actions compared to random events. Random threats can further be subdivided into a number of smaller groups, such as natural events, technical failure, human error, and so forth.

Examples of threats are

- Physical damage, such as fire, water damage, and pollution

- Natural events, such as earthquake, flooding, climatic conditions

- Loss of services, such as air-conditioning, power, and telecommunication

- Compromise of information, through, for example, eavesdropping, theft of equipment, disclosure

- Technical failures

- Deliberate threats, such as computer criminals, terrorists, industrial espionage, or insiders

The threat identification typically starts by identifying which generic threats are relevant and possible, and then these are developed into more specific threats for the specific system being considered. For example, could *fire* be identified as a generic threat, with *fire in the server room* being a more specific threat.

13.3.5   Identify Events and Causes and Estimate Likelihood

By combining threats and vulnerabilities, it is possible to identify events that may cause adverse effects and their causes. This is initially a qualitative description. Each event can be described as follows:

- A brief specification of the event and the asset or assets which may be lost or damaged in the event. An example could be *Integrity of the client database damaged*. In this example, the event is *damaged integrity* while the asset is *client database*.

- Description of the combination of threat and vulnerability which may be the cause of the event, for example, the threat being *Disk failure* and the vulnerability being *Inadequate backup routines*.

- There may be several combinations of threats and vulnerabilities that may lead to a given event. Relevant combinations should be identified to have a complete picture of what may cause the event.

The estimation of the likelihood that events will occur (the probability of events) can be done qualitatively or quantitatively, depending on the choice of method and the degree of detail in the analysis. In general terms, the estimation of likelihood can be based on:

- Historical data for similar events

- Models describing the causes of the events and the likelihood of different causes occurring

- Expert opinion

13.3.6   Identify and Estimate Consequences

The type of consequences that is considered in the risk assessment is defined as part of establishing the context for the analysis. If several types or dimensions of consequences are considered in the analysis, the consequence for each type needs to be determined for each event. For example, fire may have consequences both for provision of services, loss of data, economy, and people, and each of these consequences needs to be described for the event.

In addition to this, the extent of the consequences of an event may vary significantly in many cases. Considering the fire example above, the economical consequences may vary greatly, depending on the magnitude of the fire. Depending on the detailed circumstances of what is happening, the consequences may possibly vary from negligible to catastrophic. There are two ways to handle this situation when the consequences are to be estimated:

- Either, the event should be specified more in detail and possibly divided into several, more specific events where the consequences are more well-defined for each event. For example, *fire* may be made more specific by stating the event as *Small fire extinguished quickly.*

- Alternatively, the range of consequences can be described and the conditions under which the different severity of consequence occurs should be described. In this latter case, the conditional probability of experiencing the different consequences needs to be specified.

13.3.7   Estimate Risk Level

Estimation of the risk level essentially covers the combination of probability and consequences in a suitable way. The estimation of level of risk will depend on the method chosen for performing the risk assessment and the level of detail in the analysis:

- Qualitative analysis: The likelihood and consequences are only described qualitatively and the risk will also be expressed qualitatively. No ranking or comparison of the individual risks is performed except purely on a qualitative basis.

- Semi-quantitative analysis: By this we mean methods whereby the likelihood and the consequences are classified in categories, for example, *high probability, medium probability*, and *low probability* and corresponding categories for consequences. This means that the risk associated with each event can be classified by combining likelihood and consequence and a coarse ranking of the risks can be made.

- Quantitative analysis: Both likelihood and consequence is quantified and risk is expressed in terms of statistically expected consequence (or loss). This enables more detailed analysis and the results can also be added together to calculate a total risk for the system being considered.

The choice of method will also have to match the risk evaluation criteria which have been specified.

The resources required to estimate risk will usually be larger for a quantitative analysis compared to a qualitative analysis and the effort required therefore needs to be balanced against the additional decision support provided by a more detailed analysis.

13.3.8 Risk Evaluation

The risk evaluation is a comparison of the estimated risk level with the risk evaluation criteria that have been specified for the risk assessment process. In the risk evaluation, it may also be useful to introduce discussion of the uncertainty in the results. Risk estimation is often associated with significant uncertainty and the conclusions that can be drawn from the risk evaluation will consequently also be uncertain. This uncertainty should be discussed and presented as part of the basis for making decisions about risk treatment.

If the estimated risk level is above the decision criteria, the conclusion will normally be that risk treatment is required. However, it may also be concluded that a higher risk is accepted without risk treatment. This decision should be made by the management responsible for the system being analyzed.

Different types of decision criteria may be applied and the process of risk evaluation will depend on the type of criteria. This is further discussed in the following section.

13.3.9 Risk Treatment

If the conclusion from the risk evaluation is that risk treatment is required, a systematic approach for identifying and evaluating potential controls should be applied:

- Identify relevant controls

- Evaluate the effect of the controls in reducing risk

- Determine the cost of implementing the controls and other negative effects

- Rank the controls with respect to risk reducing effect versus cost and negative effects

The identification of controls should be based on a systematic review of threats and vulnerabilities and how risk can be reduced. A basis for identification of controls is provided in ISO/IEC 27001/2. When considering controls, it is useful to prioritize according to the following structure:

- Controls removing or reducing threats should be given highest priority

- Controls reducing vulnerabilities are second priority

- Controls reducing consequences are third priority

It may also be possible to reduce risk by transferring it to other parties, for example, through insurance.

There are several factors to consider when evaluating whether to implement a control

- How effective is the control in reducing risk? Will it have an impact only on one specific risk or a wider scope?

- How reliable is the control? How likely is it that the control will function as specified, when required?

- Are there direct negative effects of the control themselves? Will they impair operation, reduce efficiency, and so forth?

- What is the cost of implementing the control, both in terms of initial investment costs and operating costs?

  Some examples of principles for evaluating risk treatment options are

- Critera for changes in risk: Comparison of alternative risk-reducing measures to see how efficient they are in reducing risk. Select measures with the highest risk-reducing effect.

- Equality of risk: Measures are introduced to reduce risk to same or similar levels for different systems or groups of persons.

- Cost-effectiveness: The cost per unit of risk reduction is calculated, and this is used to determine the cost of implementing a measure versus the benefit (cost-benefit analysis).

- ALARP (As Low As Reasonably Practicable): Risk is reduced to a level where the cost of reducing risk further is grossly disproportionate to the benefit gained in terms of reduction in risk.

Cost-benefit analysis is often seen as a good option for deciding on implementation of risk-reducing measures. However, for events with extreme consequences (that we want to avoid more or less at any cost) but with low probability, cost-benefit analysis will often give as a conclusion that risk reducing measures are not required. In such circumstances, it may be more useful to look at consequences only and decide on risk treatment based on that.

When appropriate controls have been identified and implemented, the risk is accepted. The risk that remains after controls have been implemented is often called *residual risk*.

13.4 Summary

In this chapter, we have looked at the risk assessment process and illustrated this process in generic terms. A wide range of generic methods for use in risk assessment exists and further description can be found in Rausand [7] and Aven [5]. A number of specific methods for use in information security applications are also developed.

Risk assessment provides a structured framework for assessing risk and providing decision support when identifying and evaluating the need for risk reduction. However, there are also weaknesses in applying risk assessment uncritically.

In particular, it is noted that risk is an expression of a statistically expected consequence. This means that events with low probability will always have a low risk, even if the consequences are extremely high. An event that may have catastrophic consequences if it occurs may therefore still come out with a low or medium risk. In some cases, the consequence itself may be unacceptable, even if the probability is low.

If this is the case, the risk assessment should be supplemented with alternative approaches, for example, by looking at specific scenarios, how these may unfold, and what measures are in place to prevent these from happening. Focusing on the consequences only and how to prevent the worst-case consequences from occurring may also be an approach that may be chosen.

It is also important to note that the uncertainty associated with the estimation of likelihood in many cases can be very large. The basis for risk estimates should therefore be known and results should be used with caution, recognizing the uncertainty. There are a number of sources of uncertainty:

- Lack of data or data which are not fully relevant

- Assumptions and simplifications in the identification of events and causes

- Uncertainty about future developments and how this may affect risk

- Subjective evaluation by the analysis team performing the risk assessment

However, in many cases, risk assessment will provide a structured and systematic way of identifying and assessing the hazards and will provide a good basis for prioritizing and making decisions about how to reduce risk further.

13.5   Further Reading and Web Sites

Two very simple guides to risk assessment are

- *Five Steps to Risk Assessment*, Booklet INDG163, Health and Safety Executive, London (2006). http://www.hse.gov.uk/

- *Hazard Identification, Risk Assessment and Risk Control*, Major Industrial Hazards Advisory Paper No. 3, Department of Urban and Transport Planning, New South Wales, Australia (2003). http://www.planning.nsw.gov.au/

A comprehensive introduction to the topic is given in the book

- *Risk Assessment: Theory, Methods, and Applications*, M. Rausand, Wiley, 2011.

Another guide, although more technical, has been published by NASA

- *Probabilistic Risk Assessment Procedures Guide for NASA Managers and Practitioners*, Stamatelos et al., Technical Report, 2002. `http://www.hq.nasa.gov/office/codeq/`

Bibliography

[1] ISO/IEC 27005 information technology–security techniques–information security risk management. Technical report, ISO, June 2008.

[2] ISO 31000: Risk management–principles and guidelines, 2009.

[3] ISO/IEC guide 73: Risk management–vocabulary–guidelines for use in standards, 2002.

[4] ISO/IEC 13335-1: Information technology–security techniques–management of information and communications technology security–part 1: Concepts and models for information and communications technology security management, 2004.

[5] T. Aven. *Risk Analysis-Assessing Uncertainties beyond Expected Values and Probabilities.* Wiley, 2008.

[6] ISO/IEC 27001:2005 information technology–security techniques - information security management systems–requirements. Technical report, ISO, 2005.

[7] M. Rausand. *Risk Assessment: Theory, Methods and Applications.* Wiley, 2011.

14

Information Security Management—From Regulations to End Users

E. Albrechtsen and J. Hovden
Department of Industrial Economy and Technology Management
Norwegian University of Science and Technology

CONTENTS

Information security is not just about technology. High level of information security is obtained by the interplay between technology, organizations, and humans. Of course, the technology itself is the basis for information security, but it has to accommodate human, organizational, and societal needs to be successful. It is part of a socio-technical dynamic system governed and controlled by laws and regulations, standards, guidelines, and norms for informal behavior achieved by education and experience. A web of actors are involved at all levels, from governmental and private agencies, enterprises, and down to the individual user at workplaces. The threats and security risks have their origin in both technical and human factors. For the human factors, we divide between accidental events, such as people violating information security by carelessness, ignorance, or misunderstanding, and, on the other hand, deliberate actions motivated by malicious intentions. We need methods for information security risk identification, assessment, and evaluation for the protection of information security performance.

The aims of this chapter are to describe a framework for handling information security risks encompassing the context of norms given in regulations and standards, methods for risk assessment and evaluation as input to decision making, remedial actions and improvements. The last sections of the chapter are about social–science based organizational perspectives on information security management, that is from technical–administrative approaches to human resources viewpoints, and ending with cultural aspects of information security management.

<hr style="border-top: 4px solid black; width: 20%;" />

14.1   A Risk Governance Framework Applied to Information Security

Governance is a wider concept than management. On a societal scale governance describes structures, processes for collective decision-making involving governmental and non–governmental actors [1]. There has been a move from pure governmental regulations to governance meaning that public regulations are substituted or supplemented by auditing regimes enforcing self-regulating organizations and private- and public-based collaboration and institutions creating new forms of fragmented regulations [2]. It is part of the so-called New Public Management trends over the past 20 years based on a critique of the static and bureaucratic approaches to governmental regulations and control and represents a response to the needs for regulatory adaption to the dynamics of fast technological changes and shifts in organizational forms in terms of deregulations and privatization of critical infrastructures, and so forth [3].

Corporate governance is based on principles of risk assessment, remediation, reporting, and accountability. The increased storage and distribution of electronic information that needs to be secure creates new challenges for information security governance at all levels. *Risk governance* refers to the actions, processes, laws, traditions, and institutions by which authority is exercised and decisions are taken and implemented. It is applicable at all levels from society and regulatory institutions, to corporations, companies, and organizational units; i.e.the concept of risk governance can be used within an organization, between organizations within a nation, and internationally, i.e. vertical governance. Horizontally risk governance is about the interfaces between governmental agencies, industrial domains, science, and civil society such as NGOs, media, etc.

The concept of *risk* has many definitions [4] (see Chapter 13). A generic definition is given by Ortwin Renn [5]: Risk is an uncertain (positive or negative) consequence of an event or an activity with respect to something that humans value. In addition, a new concept of risk labeled *systemic risk* by OECD [6] is of great relevance for information security management. Unwanted events are seen as nonlinear phenomena with interdependencies between risks that emerge in complex and dynamic, systems resulting in a broad range of social, financial, and economic consequences [7].

The World Economic Forum estimated in 2008 that there is 10–20% probability of a major critical information infrastructure breakdown in the next 10 years, with a potential global economic cost of approximately 250 billion US $. An EU-based project known as MS3i (Messaging Standard for Sharing Security Information) (www.ms3i.eu/rome) is developing a management messaging standard that specifies the requirements, in terms of policies, processes, and controls, for implementing, operating, maintaining, and improving sharing security related information. *Trust* is a prerequisite for effective sharing, trust in that the recipient will not abuse the information received and trust in the provider of the information that it is validated and consequently actionable. This may sound obvious, but it is a common experience that it is a challenge to achieve and maintain such a two-way trust. Trust in the institutions and actors ensuring information security is vital for security performance and may reduce the needs for strict control regimes at all levels.

The aims of information security governance have traditionally been divided into three key elements:

1. Confidentiality

2. Integrity

3. Availability

1 and 3 are sometimes in conflict and may be weighted differently depending on the context and regulatory requirements. *Information security governance* is about addressing obligations, accountability, and responsibility among actors and stakeholders of all levels and layers as discussed by Gurpreet Dhillon [21]. We can interpret information security governance as encompassing *protection of information systems,* thus it represents an extended understanding of information security and computer security, including use of information and information technology. Protection of information systems address:

- information and data

- how information and data are processed, saved, and distributed

- how information technology influence and support organizational processes and communication

In light of this, information security governance must be understood in a broader scope than technological and administrative aspects. We can thus understand information security management as the total of activities conducted in a more or less coordinated way to control threats and vulnerabilities, both administrating routine tasks (formal) (e.g. implementation and follow-up of measures: policies, plans, guidelines, risk analysis, training programs) and guiding organizational processes (informal) (e.g. engagement of users, employee participation, decision-making).

We will look at information security governance within a generic framework of *risk governance.* Best practice in risk governance encompasses the principles for governance within the processes of risk identification, assessment, management and communication and includes criteria such as effectiveness, accountability, efficiency, fairness and social and ethical acceptability [5]. A number of criteria for evaluating risk governance are proposed:

- *Effectiveness* (Were the goals of risk management accomplished or are they likely to be accomplished?)

- *Efficiency* (Are the management measures cost-effective?)

- *Legality* (Are the risk measurement measures compatible with legal prescriptions and national/international laws?)

- *Legitimacy* (Are the management measures based on due process and publicly accepted procedures?)

- *Accountability* (Are all responsibilities for risk management and liability clear and unambiguous?)

- *Fairness* (Is the risk-benefit distribution considered fair and just?)

- *Acceptance* (Are the measures approved by the main stakeholders and the public at large?)

- *Acceptability* (Are the measures compatible with ethical and moral standards?)

- *Sustainability* (Are the measures in line with the goals of sustainable development?)

The five core risk governance processes are according to Ortwin Renn [5] (see Figure 14.1). The five risk governance in the figure processes will be elaborated and discussed throughout the chapter. But risk governance goes much further: It is about organizational capacity (assets, skills, capabilities), a network of actors (politicians, regulators, business, NGOs, media, and the public). It is also about general contextual factors and conditions such as social climate and the political and regulatory culture [4].

The levels and layers within an organization are stressed by a number of external forces and counterforces in their coping with risk problems. The main contextual stressors influencing priorities and incentives of decision making in the formation of risk governance mechanisms of adaptation to change are

- Changing political climate and public awareness

- Changing market conditions and financial pressure

- Changing competence and levels of education

- Fast pace of technological change

- New regulatory requirements

Change has now become the everyday reality of business and work. This atmosphere of constant change presents challenges to the management of risk in business and industry and to the functioning of control systems and regulations of risk prescribed by government.

The ICT risk governance system of an organization, a company, an so forth, has to relate to

- Laws and regulations, authorities, and regulatory systems

- Traditions and practices, cultural and professional factors

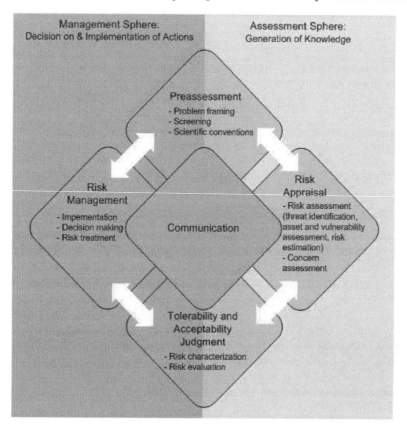

FIGURE 14.1
Risk governance framework based on Orwin Renn's book *Risk Governance*. [5].

- Industrial norms and certification

- Demands from stockholders

- Demands from customers and clients, – the market

- Demands from mass media and public opinion

Figure 14.2 models risk management in a dynamic society [9]. It shows that many levels of politicians, managers, and work planners are involved in the control of hazards and threats by means of laws, rules, and instruction. At the bottom of the model, one finds sharp-end practice, e.g. the information security performance of users without management responsibility and no information security expertise. At the top of the model, society seeks control through the legal system, by laws, and by regulations.

The public authorities regulate information security by; laws, by inspections, advisory services, and stimulation. At the same time, they are dependent

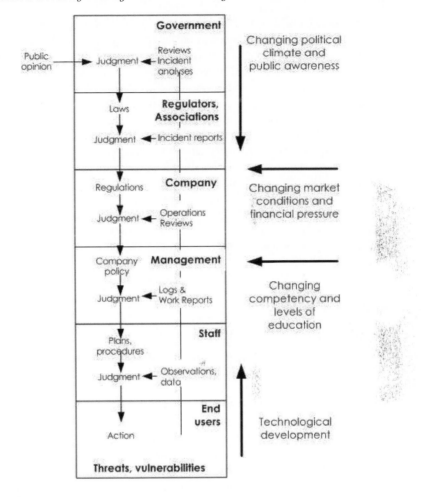

FIGURE 14.2

The socio-technical system involved in risk management in a dynamic society. Adapted from Rasmussen [9].

on input from the lower parts of the model to decide how they should regulate information security. The rules and regulations have to be interpreted in the context of a particular company and implemented by means such as policies, plans, and measures. This will directly influence security practices at the sharp end. This top-down approach shows that regulations of public authorities frame how companies organize their information security work, and thus indirectly influences user performance. At the same time, the model shows that bottom-up approaches are also essential. Higher levels need input on actual performance at the lower levels to adjust and implement security means and measures to cope with the actual threats.

The model also illustrates how environmental dynamics in society influence information security work at all levels in society. Technological change is of course an essential dynamic of information security: use of new software and hardware; new vulnerabilities in software; trends of use (e.g. Facebook); converging technologies; and coupling of systems. Differences in competency are also creating changes, in particular the difference in experience, knowledge and skills between old and young employees. Soon a new generation of workers, who have used the internet and computers from preschool age will enter the work market. What security challenges do this group of young employees represent? Market conditions and financial pressures also generate environmental stressors: for example, technology-driven organizational development and automation but also malicious acts such as industrial espionage. Public awareness and the political climate also influence risk management in society, for example by emphasizing terrorism but also on vulnerabilities in technology, regarding for example, air traffic control or the power supply.

14.2   Regulations and Control

Risk problems are a political subject, as politics embraces all issues and societal processes concerning the means and ends for disbursing benefits and burdens among citizens. The German sociologist Ulrich Beck argues that the distribution of risks will be the main political issue in the future [10]. He argues that risk issues related to justice and participative principles may dominate the future political agenda on a global basis.

Regulation is addressed primarily as an activity that operates externally to a technology, business, industry, transport system, or other activity. Regulations are imposed on the activity by parties outside, often the government on behalf of society. The reason is concern or dissatisfaction with the risks of the activity. Regulation is therefore about control or restriction. The motives for doing this can be sought in protection of weaker groups or in vital interests of the state, in preventing a waste of economic resources, and in guaranteeing a fair arena for market competition, for example EU framework regulations.[1]

Information security regulation is about the regulation of risks. In this respect, it differs fundamentally from many other kinds of regulation. The risk of losses is ever present as we go about our lives – at work, as we travel, as we produce or consume. All that can reasonably be required of those who regulate and control business and workplaces, critical infrastructures, products and services is that they *minimise* the risks.

The main societal/governmental means for influencing the contents and performance of information security management by companies are laws and

---

[1]See http://cordis.europa.eu/fp7/ict/security/policyen.html

regulations, surveillance, direct controls and sanctions, rewards and support, providing information, education and training. The authorities are also actors in collaboration with employers and employees organizations, and other stakeholders and interest organizations on the basis of voluntarily achieving mutual consent on how to deal with information security issues.

The institutional arrangements of regulatory information security regimes vary between countries, but their functions are quite similar. In Norway, there are several regulators concerned with information security [11]. Some of these authorities are covering all business domains, for example the Norwegian National Security Authority, the Data Inspectorate, and the Norwegian Post and Telecommunications Authority. Additionally, there are sector-based authorities regulating information security, for example the Financial Supervisory Authority of Norway, the Directorate for Health and Social Affairs, and the Petroleum Safety Authority Norway. These public authorities regulate information security by laws, inspections, advisory services, and stimulation.

Market demands are playing an important role for increased attention and motivation for improving companies' information security management systems. Bad information security performance may bring companies out of business by losing customers and contracts. In a framework of a market economy, certification of systems, processes, and products, combined with insurance of losses and liability claims have partly substituted direct control and regulations by government.

External actors and stakeholders in regulating a company's risk management:

- Governmental institutions; local, national, supranational

- Trade unions

- Financial organizations and trade institutions; stockholders

- Mass media

- Customers

- Interest groups; consumer groups, internet groupings, and so forth

- Certification bodies

- Insurance and liability

- Consultants and contractors

- Professional organizations

The establishment of national risk regulatory regimes dates back to the 19th century when governments recognized their mortal obligations for health and safety for their citizens. Insurance institutions also became actors in loss prevention and risk distribution. Safety regulations related to industrial safety and major accidents became the strongest societal constraints on business and industry. In this perspective, ICT regulatory regimes on information security activities are a rather new. The information technology is developing at a fast pace. Therefore, the regulations, controls, and enforcement in this area are rather immature and relatively little harmonized and coordinated compared to the regulations of physical risks. But ICT is an integral part of all critical infrastructures in the modern society, in industrial production, transportation and so forth. As such, ICT has potential for causing or contributing to injuries and damages. Information security requirements are therefore partly encompassed in the general risk regulations for each of these specific business domains. The organizing of the information security regulatory regimes varies a lot between countries.

The development of Information security regulations has to a large extent been driven by experts with a military or police background in combination with specialized ICT experts. This created a regulatory culture that differed from other risk regulations in society [12]. Information security regulations try to some extent adapt and harmonize with more mature risk regulations in industry and transportation:

- From *Need to know* doctrines to *Right to know – about facts and values regarding risk communication:* Traditionally have requirements on "confidentiality" been the main focus in information security management, which influence the willingness to be open about threats, incidents, and the risk management system. Openness about information security problems and their handling can increase the systems vulnerability for attacks. This represent a dilemma in the interface with other risk management systems based on "right to know" principles.

- *From end result control (and punishment) to a problem-solving approach/goal-setting:* The first approach is reactive and will punish not succeeding, independent of how hard you try, and rewarding a rule following bureaucratic approach. The other approach is proactive and analytical, facilitating technical and organizational innovation, including adaptation to new risks and change forces.

- *From detailed, prescriptive regulations to functional requirements:* The "old" approach requires company access to rules and standards, looking for documentation, and deviation control. The second approach requires access to modern management principles, analytical methods and techniques, searching for integrated socio-technical means for improvements. The second approach also requires higher competence in information security management system's design, running, and auditing.

- *From reactive control and prioritization to risk analysis system based approaches and requirements:* This summarize the description of main differences in approaches to risk regulation. On the one hand, control by specific requirements, rules, and standards for how to achieve acceptable security results. On the other hand, a system-oriented holistic approach of functional and dynamic requirements on performance, and not on how to achieve the results.

Control-based prescriptive regulations are typical for the United States, – and NATO, whereas EU is more laidback and includes some aspects of a systems-oriented approach. In the preparations of the Norwegian Security Law,[2] they had some intentions of applying some risk-based functional requirements, but ended with more than 1000 specific requirements and recommendations based on NATO guidelines. This law is mainly about the security of the state/nation, which is different from protecting assets for companies. But it is not easy to separate security regulations of critical infrastructures and critical societal functions from implications and consequences at lower levels, for example the financial sector. In some domains, security subgoals may be in conflict. For national security, confidentiality is usually the main concern and overruling the others. In the health services integrity, availability and personal privacy are priorities. The last one is sometimes in conflict with the two first concerns.

National security issues and some critical infrastructure risk and vulnerability problems cannot be handed over to be solved by market forces but enforced by governmental regulations, that is, a need for "hard laws". However, information security management of business activities are to a large extent in most countries regulated by "soft laws" and nongovernmental bodies or so-called New Public Management arrangements, and so forth:

- Private/public-based institutions monitoring and warning on information security threats

- Authorization of standardization and certification institutions

- Product stewardship ("responsible care") by business and industry

- Supranational agreements on harmonization and guidelines, for example, OECD Guidelines for security of information systems

A widely used reference standard is ISO27001/2 "Code of practice for information security management" which is presented and discussed in section 14.3 of this chapter.

It is generally recognized that private end users, for example, small companies, need help through detailed regulations and support, whereas professional, advanced users should take more responsibility themselves within a framework of functional meta-regulations. By that the regulatory control authorities can focus on auditing that the companies are controlling their own compliance with general risk governance principles, for example, Goal setting,

---

[2]http://www.lovdata.no/all/nl-19980320-010.html

Management responsibility, Systems for risk assessment, Systems for deviation control, Action plans, Documentation, and System audits, Flexible and locally anchored solutions.

Orwin Renn describes alternative risk management approaches to be used, depending on the characteristics of the risks and threats involved. The two most relevant for information security are

- *Risk-based information security regulations and management:*

  - Emphasis on formal risk analysis and assessments
  - Reduction of exposure and/or probabilities
  - Risk management according to expected values on risks and benefits, eventually combined with a risk aversion factor for a stronger emphasis on high consequences than on low probabilities
  - Reliance on inspections and routine controls
  - Modeling of variability for coping with (intraindividual) variation and changing contextual conditions.

- *Resilience-based information security regulations and management:*

  - Emphasis on cross-disciplinary knowledge and investigations
  - Containment of application (in time and space)
  - Constant monitoring
  - Redundancy and diversity in information security design
  - (Strict) liability
  - No tolerance policy for risk control
  - In extreme cases: prohibition

14.3   Information Security Management

14.3.1   Formal and Informal

It is simple and banal, but sometimes forgotten: You cannot manage anything without knowledge and understanding of what you are managing. The most sophisticated and systematic management system does not help without some generic knowledge of the functioning of organizations and of organizational life, plus of course the technology of the production processes involved. Management is to secure the functioning of an organization. If we wish to manage and improve the information security conditions we face in organizations, it is necessary to study how organizations function; how humans think and act; how technology can provide protection; and what risks the system face. Information security has traditionally been technology-oriented discipline, with a large number of available technological security solutions [14] [15]. However, by the widespread use of computers at both work and home; the increased connectivity and access to information; the communication channels available by information technology; convergence of technology; and the utilization of technology in new organizational forms and ways of organizing work, non-technological aspects of information security now must be considered in addition to the technological aspects. As a result, information security must be treated as a socio-technical system.[3]

A socio-technical information security system is created by elements of all information security processes and the interplay between these elements [16]:

- technological solutions

- policies, guidelines, and instructions for individual and organizational behavior

- methods and tools

- the role and responsibilities of information security professionals

- individuals' behavior, awareness, expectations, and experiences

- collective norms and values

- interactions and relations between individuals and groups

These related factors produce a web of technological and nontechnological elements that per definition should create a secure, reliable, and available information system. Information security should thus be addressed holistically by interdisciplinary efforts.

The socio-technical mind-set was created long before information security began to play an important role in organizations and society. It has its roots in the 1950s, when studies showed that technical and social systems are closely linked [17]. The school argues that incompatibility between demands created by technology and what is beneficial for workers' situation do not create improved performance. This study was followed by several socio-technical studies, which have had a major influence on participative approaches to organizational development and safety management in the Scandinavian countries. Socio-technical theory emphasizes the development of humane working conditions and realizes ideas of participation and democracy. Socio-technical theory proposes a number of different principles for handling technical and social aspects of an organization as inseparable elements [17], which all are relevant for

---

[3]An MTO-perspective (Man-Technology-Organization) is often used as a synonym for socio-technical systems.

managing and understanding information security: joint optimization of technology and social organization; man as complementary to the machine; man as a resource to be developed; optimum task grouping; self-regulation subsystems; redundancy of functions; at organization chart, participative style; collaboration; purposes of member, organizations, and society; commitment; and innovation.

Following this lines of thought, information security management concerns a broader scope than just technology. Information security expert Schneier [18] illustrates this neatly by stating:

> If you think technology can solve your security problems, then you don't understand the problems and you don't understand the technology.

As a result, information security can be understood as *protection of information systems*, thus an extended understanding of information security and computer security, including use of information and information technology [22]. Protection of information systems deal with protection of (1) data and information; (2) processes were how information and data are processed, saved, and distributed; (3) how this supports organizational processes and communication supported by the latter two.

In light of this, information security management must be understood in a broader scope than technological aspects. We can thus understand information security management as the total of activities conducted in a more or less coordinated way to control threats and vulnerabilities, both administrating routine tasks (formal) (e.g., implementation and follow-up of measures: policies, plans, guidelines, risk analysis, training programs) and guiding organizational processes (informal) (e.g., engagement of users, employee participation, decision making), 14.3

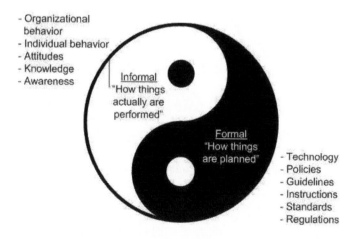

FIGURE 14.3
Formal and informal information security management.

In the formal part, we find technological solutions and structural organizational elements (e.g., policies, guidelines, standards) describing planned and expected organizational and individual behavior. *Traditionally, information security management has been dominated by this formal and structural perspective on organizations.* Although it is required to add an informal approach, it does not imply that current hard approaches should be rejected:

- Information security is and still will be a technological discipline. Without technological information security focus and solutions, every information system will collapse.

- The formal, documented information security system must still be in place without it, systematic and informal management would not be possible. Policies, plans, guidelines, and standards will function as a guiding star for the information security work.

However, we also need a complementary informal approach. While the formal part is mainly related to how things are planned to be performed, the informal part is concerned with how things really are done in practice. It is thus about how information security is preserved in daily human and organizational activities. In an informal approach to information security, we find elements of a socio-technical information security system related to the human-resource, political, and symbolic perspectives of an organization such as individual behavior, awareness, and knowledge; information security culture; and decision-making and power.

Based on Mintzberg [19], the rest of the section is divided into two main parts: *a formal view on organizations* and *an organic, informal view*. These two views are further elaborated based to Bolman and Deal's [21] perspectives on organizations. There are several valid perspectives, frames, or images of organizations. Each is interesting and significant, but each gets at only part of the truth. Each frame or perspective provides a different way of interpreting events and actions, and each implies a very different focus on effective management. Together, they offer a comprehensive view [21], leading to a greater freedom of choice for remedial measures. Separating the section in this manner makes it possible to utilize important theories within the social sciences for different views of the organization as well as show the interplay between the formal and informal parts of an organization.

14.3.2 Formal Approaches to Information Security Management

First, let us look at a general definition of organizations: *An organization is defined as two or more people working together cooperatively within identifiable boundaries to accomplish a common goal or objective. Implicit in this definition are several important ideas: Organizations are made up of people (i.e, members); organizations divide labor among members and organizations pursue shared goals or objectives.*

The formal view on organizations can be described by using Bolman and Deal's [21] structural organizational frame, which is based on the following set of core assumptions:

- Organizations exist primarily to accomplish established goals.

- For any organization, there is a structure appropriate to the goals, the environment, the technology, and the participants.

- Organizations work most effectively when turbulence and personal preferences of participants are constrained by norms of rationality.

- Specialization permits higher levels of individual expertise and performance.

- Co-ordination and control are accomplished best through the existence of authority and impersonal rules.

- Structures can be systematically designed and implemented.

- Organizational problems usually reflect an inappropriate structure, and these can be resolved through redesign and reorganization.

The approaches to information security management described in standards and in literature have been dominated by a formal view of organizations [14], [23], [24]. It is not only the information security management discipline that has its origin within this framework. In fact, most of the general literature on management and organizations has it roots in this view on organizations.

ISO/IEC 27001/2 Information Security Management

Important steering documents for regulations and information security management in organizations are the ISO/IEC standards 27001 "Information technology: security techniques: information security management systems: requirements" and the 27002 "Information technology – Security techniques – Code of practice for information security management" [24] (formerly known as ISO17799 and BS7799-1/2). These standards provide a formal set of specifications for a information security system. The standards are part of the 27000-series, which also includes, measurement and metrics; risk management approach; and a guide to certification. It is possible for companies to get their information security management system be certified by a third party according to the requirements in the standard.

The 27001/2 standards give an indication of what formal information security controls should be in place:

- Risk assessment

- Security policy

- Security organization – governance of information security

- Asset classification and control

- Personnel security

- Physical and environmental security

- Communications and operations management

- Access control

- Systems acquisition, development, and maintenance

- Information security incident management

- Business continuity management

- Compliance with legal requirements

These security controls are found within a structural framework of organizations. They offer structures and standardized solutions that should fit the goal and technology of the organization.

One problem can arise when using management systems described in standards. According to contingency theories, it is important to adjust proposed controls to the actual context of organizations and adapt the formal systems to the informal systems of an organization. The *contingency approach* states that there is no one best or most appropriate way for all organizations to structure or organize. The best-fitting structure depends on the context that the organization faces. The contingencies are usually grouped as follows:

- the organization's environment (markets, regulations, etc.)

- (core) technology (mass or batch production, primary processes)

- goals or objectives (business, social, safety, security, etc.)

- size (specialization and coordination aspects, formal versus informal structures)

- culture (matching the structure)

From Formal Approaches to Informal Approaches to Information Security Management

Partly as a consequence of information technology, new ways of organizing work has emerged. Bureaucratic, predictable organizations are by deregulations, privatization, outsourcing, and so forth, been substituted by for example, virtual organizations, partnerships, home offices, and mobile workers (and equipment). We also see an improved level of skills and knowledge on use of information technology, particularly when new generations of workers enter the market. This improved level of skills will probably lead to both positive (e.g, improved awareness) and negative effects (e.g, knowledge on how to misuse systems). The development creates new challenges for information security management, such as new work forms; distributed organizations; technology development; mobile equipment; new generations of workers, for example, "the Playstation generation"; and increased complexity and dynamics within and outside organizations. In addition, there are dynamics in society that information security management in organizations must deal with, such as market and financial conditions, public awareness, and the political climate. Furthermore, we know that information security threats and vulnerabilities are an emergent challenge characterized by dynamic, complex, and uncertain problems.

These developments within and outside organizations require new approaches to management of information security. Some new ways to address information security management has been proposed: the RITE principle [26]; development of information security culture; and participative information security management [16]. The latter two will be dealt with in the subsection on informal perspectives on organizations, while the RITE principle will be described here.

From CIA to RITE

Information security has traditionally been concerned with the preservations of confidentiality, integrity, and availability (CIA). However, the new organizational complexities listed above create challenges for information security management, which these three dimensions cannot cover sufficiently.

Confidentiality is to restrict data access to those who are authorized. However, in modern organizations information is the very lifeblood of organizations with an intense sharing of information among members and organizations. The production of an organization depends on protection of information systems; however, the challenge is that information should be dynamic, updated, and shared, making information available to the many, not the few. Preserving confidentiality thus can be an obstacle for modern ways of organizing, particularly in knowledge organizations. Preservation of integrity also faces new challenges in modern organizations. Integrity is based on the idea that information should not be changed or altered; however, in modern organizations information should be updated, relevant, and complete. Regarding availability, system failure is an organizational security issue. The availability notion thus becomes more important in new organizations than previously, where confidentiality has had the main focus, which is in contrast to previous information security approaches, where availability has been the notion paid least attention to.

Many organizations today are geographically distributed, and workers are equipped with mobile equipment. CIA, on the other hand, is based on an assumption that information and data are stored at one location. Paradoxically, risk becomes distributed while information security management remains centralized. The organizational development, on the other hand, implies a need for distributed responsibility.

To handle challenges for information security management in modern organizations, Dhillon and Backhouse [25] suggest four alternative principles to CIA, called RITE:

- Responsibility and knowledge of roles. Members must understand their responsibilities and roles in information security organizations.

- Integrity as a member of an organization.

- Trust (as distinct from control). Less control and monitoring and more emphasis on self-control and responsibility in a mutual system of trust between users and managers.

- Ethicality (as opposed to rules). The ethical content of informal norms and behavior.

14.3.3   Informal Aspects of Information Security Management

Subsection 14.3.2 described a formal view on information security management and organizations. In this subsection, we look at a complementary view on management and organizations. While the formal view looked at how organizations are structured; how management is planned; and how organizational and individual behavior is expected to be, the informal perspective rather looks at actual human and organizational performance; how humans are a resource in the information security system; and norms and values. In this section, we have delimited the informal view to human resources and organizational culture. With a wider scope, this organizational view also can include organizational processes such as decision making; politics and power; and information processing, which are important for understanding risk governance and risk assessment in organizations.

The Human Resource Approach: The Human Part of Information Security

The human resource frame draws from a source of research and theory built around several major assumptions:

- Organizations exist to serve human need, not the other way (the analogue is the controversy of fitting people to machines, or fitting machines to people at the workplace level).

- Organizations and people need each other. Organizations need the ideas, energy, and talent that people provide, while people need the careers, wages, security, and work opportunities that organizations provide.

- When the fit between the individual and the organization is poor, one or both will suffer.

- When the fit is good between the individual and the organization, both benefit.

There is a difference in assumptions on the human nature between the structural and the human resources frames. Whereas the structural approach tries to compensate for human failures, imperfection, unreliability, self-interests, and limited rationality, the human resource frame focuses on the positive aspects of human nature: coping, learning, adaptation, competence, responsibility, and social and caring values.

One can often see news articles giving examples of information security breaches caused by poor user behavior or insufficient awareness (it must be assumed that these incidents only represent the tip of the iceberg), for example:

- *Incautious use of e-mail:* before it was made public, a budget was accidentally sent to a newspaper rather than to the correct receiver, a public agency.[4]

- *Lost mobile equipment:* At Oslo Airport, the lost property office receives about three computers every day.[5]

- *Finger mistake:* A stockbroker accidentally typed wrong numbers and unintentionally bought stocks for NOK 40 million, which were later sold with a loss of NOK 6 million.[6]

- *Sensitive information made public available:* The Norwegian National Security Authority found both trade secrets and security-graded information in Facebook profiles of members of public agencies in Norway.[7]

These examples are mainly unintended acts. However, the accidental dimension of user-created incidents has not been addressed adequately in the information security domain [26]. On the other hand, much attention has been paid to users as "the enemy within." For example, there are security incidents

---

[4]http://www.vg.no/pub/vgart.hbs?artid=116162

[5]http://www.dagensit.no/min-it/article864669.ece

[6]http://e24.no/arkiv/article659858.ece

[7]http://pub.tv2.no/nettavisen/innenriks/ioslo/article1315514.ece

that are founded on employees being tricked. Mitnick and Simon [27] give several examples of how social engineering can be used to attack information systems;that is, hackers use social techniques to manipulate people into performing actions or give away confidential information. In a similar way, phishing attempts and Nigerian fraud approaches are based on tricking people to perform actions they should not be doing.

Furthermore, malicious acts of legal users of a system are a major threat to information security. Gordon et al. [28] show that nearly half of the reported computer crime incidents in the United States are created by insiders, for example, abuse of net access; unauthorized access to information; sabotage; theft of software or equipment and fraud. It is widely assumed that a remarkable portion of information security breaches in an organization are carried out by its own organizational members (e.g, [27], [29], [30]). This insider threat is understood as people who have been given access rights to an information system and misuse their privileges, thus violating the information security policy of the organization [31].

These examples indicate that users can be a possible threat/vulnerability for the information security level either by deliberate or accidental incidents or by being tricked to create information security breaches. Blaming users for these incidents would be to go back to the mind-set of the occupational and industrial safety discipline 20–30 years ago, when individual failures were emphasized as the main cause of many accidents [30]. Blaming the operator rather than the technology or organizational aspects has a long history in the analysis of failures and accidents. Human failure is often the first and the most common attribution when accidents occur, such as the Chernobyl catastrophe, airplane disasters, and major train accidents. Rather than giving the blame to the operator, one should ask what in the system made it easy for operators to make mistakes [31]. We can thus assume that *individual information security acts (both normal operation and when creating security breaches) are generated by various factors in technology, at the local workplace and in the organization.* This statement needs some clarifications.

First, this does not imply that we neglect that some employees have incentives to get some sort of gain by malicious acts. However, it is technological and organizational vulnerabilities that create windows of opportunities to carry out malicious acts. For example, lack of organizational information security measures (mainly lack of segregation of internal control) made it possible for Nick Leeson, a trusted general manager at Barings Banks, to exploit the substandard information security systems to do unsupervised speculative trading, thus making large personal profits that finally caused the collapse of Barings Bank, the United Kingdom's oldest investment bank in the early 1990s. Second, human behavior is by nature unreliable. Proper barriers must thus be in place to prevent information security incidents. Barriers are here understood as physical and/or nonphysical means planned to prevent, control, or mitigate undesired events. The barriers can take many forms [32], ranging from physical (prevent an action to be carried out); functional (impeding the

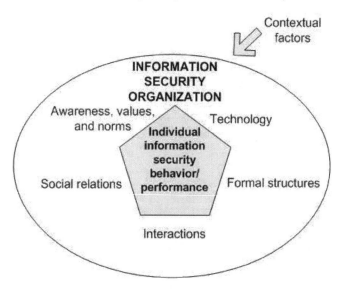

FIGURE 14.4
Individual information security performance explained by organizational aspects [15].

action to be carried out, e.g, password authentication); symbolic (interpretations required in order to act, for example, warning messages and interface layout); and incorporeal (the barriers are not physically present, but depend on the knowledge of the user in order to achieve its purpose, e.g, rules, guidelines, and security norms and values). Poor quality or lack of one or more of these barriers creates possibilities for information security breaches where human acts can be the source of ignition. Having these barriers in place is the responsibility of managers, not a user responsibility; consequently, users cannot be blamed for making accidental incidents. Third, user's information security behavior is normally preventive rather than dismal. Such normal behaviour is generated by a number of contextual factors (see Figure 14.4).

14.4 illustrates how individual information security behavior can be explained by a number of organizational aspects. The model is adapted and adjusted from Schiefloe [33].

The model illustrates that information security behavior and thus individual performance (performance is the result of the behavior, i.e, action or inaction) is influenced by a set of organizational aspects: formal systems; technology; values and knowledge in the organization; interactions; and social relations.

- *Technology* is of course an important factor for information security behavior, that is, all kinds of technological security solutions, for example, access control.

- *Formal structure* covers the formal distribution of responsibility and tasks and steering documents such as policies and instructions.

- *Interactions* concern how individuals and groups cooperate, communicate, and coordinate their actions with one another. How management is performed is an important ingredient of this dimension.

- *Social relations* are about social networks, collegial conditions, and professional divides. Keywords are trust and access to knowledge and experiences.

- *Awareness, values, and norms* both individual and shared with others play an important role and are closely related to behavior. These are important factors concerning how people interpret situations and choose their actions, thus influencing work practices and norms. The attributes are influenced and maintained by formal structures, interactions, and relations.

- *Contextual factors* influence the organizational and technological information security attributes, such as other organizational processes and requirements; technological development; legal requirements; and standards. See also 14.4.

The Human Role in Information Security: Foe and Friend

The previous section above illustrates that the human element of information security is an important threat. However, users are also one of the most important resources by preventing, detecting, and reacting to unwanted incidents. Employees might be a resource for the systematic information security efforts of an organization by simple, no time-consuming actions, such as

- locking the computer when they are absent from it

- good password etiquette

- cautious use and transportation of mobile equipment

- cautious use of e-mail and e-mail addresses

- cautious use of the internet

- cautiousness at home offices

- not using unlicensed software

- not distributing confidential, internal, sensitive, or private information to people it is not relevant for

- reporting incidents and vulnerabilities or suspicion of these

In the safety research domain, resilience engineering has emerged as an innovative and new way to think about safety. This approach argues that safety is a core value, not a commodity that can be counted – safety is revealed by the events that do not happen. A key issue here is foresight – the ability to anticipate changing shapes of risk before failure and harm occurs. This is in contrast to the traditional reactive approach driven by events that have happened. This school of thought further argues that "success belongs to organizations, groups and individuals who are resilient in the sense that they recognise, adapt to and absorb variations, changes, disturbances and surprises." Consequently, the dynamics of normal operation becomes an important loss prevention process. In addition, incidents are interpreted as an unexpected combination of normal performance variability. Viewing users as an important security resource is linked to this focus on normal operation rather than hindsight on how and why incidents occur. The bulleted actions above are normal operation and even common sense/good manners, rather than complex, time-consuming security actions, and could easily be integrated into regular work tasks.

A good question here is whether it is necessary at all for users to consider the actions bulleted above. One can answer "not" to this question, in the sense that there are technological defenses-in-depth that will prevent most security breaches to escalate if the actions above are not followed. On the other hand, the answer is yes for several reasons. First, poor quality of the actions can ignite external attacks (e.g, password in the wrong hands) or open vulnerabilities (e.g, download an unlicensed program containing malicious code). Second, many of the actions are protecting the public image of the organization (e.g, cautious handling of sensitive information). Third, reporting incidents and insecure conditions are an important principle in systematic information security management.

The belief in employees as a resource is closely linked to organizational democracy and employee participation.

Information Security Measures Directed at Users [16], [37]

The field of safety psychology, which provides basic knowledge for understanding safe and unsafe behavior, categorizes measures directed at individuals into different groups. It claims that there is a sequence of ordering between these categories for the most effective strategy for including human safety performance [35], [36].

- First, one should change the preconditions in the working environment to be satisfactory for secure behavior.

- If this is not sufficient, educate workers.

- If education is insufficient, inform employees to improve their attitudes.

- If the effect of information is not satisfactory, modify behavior by sanctions and rewards.

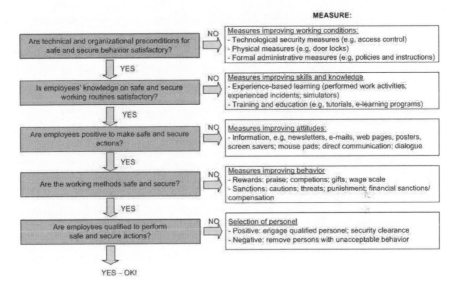

FIGURE 14.5
Information security measures directed at users [24].

- And, selection of employees is the final solution to deal with the undesired safety risks of employees.

- Relocate or dismiss unqualified employees and provide working tasks according to qualifications.

The relationships between the above-mentioned categories are illustrated in Figure 14.5. Although these strategies are developed within the industrial safety domain, the ideas are transferable into the information security field as well.

The working conditions create the environment in which employees perform their jobs. This environment consists of technological tools and formal administrative measures in addition to cultural conditions (norms, relations, and interactions between individuals).

Technological measures are of course an essential tool when it comes to influencing user performance. Security in applications, services, operation systems, kernels, and hardware creates a secure environment for employees' use of ICT systems by restricting their freedom. Technological security measures typically restrict access rights by the wellknown "need to know" principle and may also restrict the freedom by "separation of duties." Malware and intrusion detection and prevention software are expected to prevent and react to whatever improper actions users make. Computer security systems should, however, not only preserve security, they should be usable for users as well.

In addition to technological measures, technical-administrative means provide premises for individual and organizational behavior by policies, instructions, and plans that document and specify expected behavior. The main emphasis on nontechnological information security approaches has as presented in section been such technical-administrative measures.

Measures aiming at improving skills and knowledge are either experience-based learning activities or systematic training and education. The former being firsthand learning by personal experience and the latter being second-hand learning by formal education [37].

Measures directed to improve attitudes can be applied in four ways: (1) to directly change behavioral patterns; (2) to change the attitude the behavior is a result of (affection); (3) create attentiveness to security questions; and (4) make a deterrent effect [34]. Such measures can be used to improve employees' knowledge and points of view on security measures, that is, improve the security performance by making employees perceive security technology, instructions, and training programs to be positive. Voss [38] and Hubbard [39] give the following outline of information security awareness measures:

- *Notifications*: newsletters, quick notes, e-mails

- *Competitions*: contests, games, rewards

- *Arrangements*: formal presentations of security policies; guest speakers on particular subjects lunch meetings; discussion groups; Security Awareness Day or Week, movies

- *Electronical information*: web pages, intranet, screen savers

- *Public information*: posters, pamphlets, pictures and artwork, signs

- *Physical reminders*: mouse pads, tension squeeze balls, pens

There are basically two kinds of awareness campaigns: (1) society-based campaigns, which are characterized by use of experts, individual interventions, and large population groups, and communicated from authorities to single individuals; and (2) community-based campaigns, which use resources in the local community (empowerment), focus on individuals and groups, and is characterized by cross-disciplinary cooperation.

Rewards and punishment aim at controlling the frequency and form of behavior by influencing the consequences of the behavior in a positive or negative form for the relevant users. Instruments for rewards and punishment can be social tools (i.e, positive feedback, praise, competitions, warnings, punishment, and dismissal) or material/economic tools (gifts, rewards, wage systems, economical sanctions/penalties).

The measures above aim at improving personnel's qualifications and presumptions for adequate employee security performance. Selection of personnel is the opposite: people are selected to do jobs based on their qualifications, that is, positive selection. There is a strong tradition in the security field to use security clearance of personnel.

Participative Information Security Management

Participation is a means for giving the employees more opportunity to influence decisions that affect their work. Many studies of participation at work have found significant improvements in both morale and productivity, and in safety performance. Participation has been one of very few measures that has demonstrated positive effects on a number of objectives at the same time.

Employee participation has not had a strong position in the field of information security. A search in public standards and guidelines for information security reveals a very modest focus on employee participation [40]. This is in contrast to several other fields of practice with different degrees of similarities to information security that argues for worker participation, for example, safety management, technological development, and organizational development. The arguments of these fields are mainly based on a democratic mind-set regarding the right to influence working conditions; utility-driven ideas of improved ownership and motivation among workers; improved decision making and development and implementation of technological solutions; and reduced level of risk.

Participation in information security can solve information security issues such as usability and functionality issues of technology; improve information security awareness, ownership, acceptance, and motivation among employees; reduce the gap between information security experts and workers. In addition, a participative approach will ensure democracy at work, which is an important principle in, for example, Scandinavian countries.

However, one can also argue for negative consequences of a participative approach to information security. First, participation in large scales is resource demanding for the organizations. Second, the need-to-know principle has been an important strategy for ensuring confidentiality of information systems. Involving employees might jeopardize this principle. Third, by looking at users as the enemy within, one can also argue that participation is an unwanted approach as it implies that malicious employees will acquire knowledge of vulnerabilities and attack possibilities. However, a participative approach to information security does not necessarily imply contact with sensitive information. Rather it is the processes behind the participation that is important for creating improved support for decision making and comprehension of the information security practices among the security managers as well as improving awareness among users.

14.3.4 Information Security Culture

The symbolic frame is based on some basic assumptions about the nature of organizations and human behavior [20]:

- What is most important about any event is not what happened but the meaning of what happened. The meaning of an event is determined not simply by what happened but by the ways that humans interpret what happened.

- Many of the most significant events and processes in organizations are substantially ambiguous or uncertain. It is often difficult or impossible to know what happened, why it happened, or what will happen next.

- Ambiguity and uncertainty undermine rational approaches to analysis, problem solving, and decision making.

- When faced with uncertainty and ambiguity, humans create symbols to reduce the ambiguity, resolve confusion, increase predictability, and provide direction. Events themselves may remain illogical, random, fluid, and meaningless, but human symbols make them seem otherwise.

This view on organizations is closely related to the fuzzy notion organizational culture. Organizational culture is a collective representation of how people think and acts within an enterprise. The concept of culture can be divided in two parts:

- *The contents*: an invisible or latent part encompassing shared basic norms and values related to leadership, human resources, cooperation, the primary processes, risk perceptions.

- *The expressions*: a visible or manifest part encompassing goal setting, formal systems, structures, strategies, symbols, rituals, and behavior.

In a cultural perspective, the three other organizational frames represent cultural expressions. Schein [42] divides culture into three levels in the same logic as the dichotomy above:

- *Basic assumptions*: relations to environment; nature of reality, time, and space; human nature, activity, and relationships.

- *Values*: testable on the physical world or testable only by social consensus.

- *Artifacts and creations*: technology, art, visible, and audible patterns of behavior

Schein [41] argues that the term culture should be reserved for the deeper level of basic assumptions and beliefs that are shared by the members of an organization, that operate unconsciously, and that define in a basic taken-for-granted fashion an organization's view of itself and its environment. These assumptions and beliefs are learned responses to a group's problem of survival in its external environment and its problem of internal integration.

A lot of case studies on disasters and risk issues has attempted to make sense from a cultural theory perspective [42]. The analysis is mainly based on responses at the society and interinstitutional levels, but also applies to the company level. Westrum distinguishes three cultures based on the organization's response to warning signals of disasters and high-risk exposure, and to the tendency to learn:

- *Pathological*: The organization is ruled by a desire to preserve status quo: denial of signals, punish whistle-blowers, attack reputation of scientists, avoid reporting recording – an out of sight – out-of-mind attitude.

- *Calculative*: The organization plays with the rules, stays within normal wisdom, downplays signals, sugarcoats, pass of incidents as untypical, looks for scapegoats, ignores wider implications, limited scope of repair and remedial actions

- *Generative*: The organization is concerned with goals and learning. Rules are subordinate to that. It welcomes and encourages danger signals, disseminates, sees wider implications, and is positive to system changes.

The classification is partly speculative, but it is suggestive in the way it intuitively links some basic cultural features to the other frames, and especially the framework of systems theory and problem-solving models. It emphasis the role of leadership in determining culture, pleads for their physical and psychological closeness to problems to give signals of importance and to break groupthink, to design organizations for and reward upward communication of criticism, and to set a balance of production versus safety. Westrum also analyses external political and economic pressures for their tendency to promote risk taking and to suppress a generative culture.

Information Security Culture

Information security culture is a difficult and foggy concept, with many interpretations and approaches. Information security culture is a hot topic in information security work, but also one that creates confusion.

> Although many researchers have identified the importance and the need for an information security culture in organizations, few have established a clear and definitive meaning to the term security "culture" [43].

While culture is a new concept in the information security field [44], it has been around for a time in the industrial safety domain. As for information security culture, it is unclear what safety culture is and is often understood with several elements. Hale [45] suggests that the following dimensions of a culture cover many of the common interpretations of a safety culture:

- The importance to safety given by all employees, in particular top managers.

- Which aspects of safety in the broadest sense of the word are included in that concept, and how the priority is given to and felt between the different aspects.

- The involvement felt by all parts of the organization in the process of defining, prioritizing, and controlling risk

- The creative mistrust which people have in the risk control system, which means that they are always expecting new problems, or old ones in new guises and are never convinced that the safety culture or performance is ideal.

- The caring trust that all parties have in one another, that each will do its own part, but that each (including yourself) needs a watchful eye and helping hand to cope with the inevitable slips and blunders that can always be made. This leads to overlapping and shared responsibility.

- The openness in communication to talk about failures as learning experiences and to imagine and share new dangers, which leads to the reflexivity about the working of the whole risk control system.

- The belief that causes for incidents and opportunities for safety improvements should be sought not just in individual behavior, but in the interaction of many causal factors.

- The integration of safety thinking and action into all aspects of work practice, so that it is seen as an inseparable, but explicit part of the organization.

These factors apply to information security as well. What we also see from these factors is that culture is not about how individuals behave and think, but how a group of minimum two people interact. An information security culture is thus an important addition to the structural information security efforts and individual efforts. The technological, structural, individual and cultural factors must be adapted to each other; it is particularly necessary that the formal technical-administrative systems is adjusted to the informal organizational contexts.

14.4   Further Reading and Web Sites

For more details on risk governance, we recommend the text book by Ortwin Renn *Risk Governance. Coping with Uncertainty in a Complex World* [5]. The International Risk Governance Council's webpages www.irgc.org provide guidance of and several examples of application of a risk governance framework.

The two books by Gurpreet Dhillon, *Information Security Management. Global Challenges in the New Millennium* [21] and *Principles Of Information Systems Security. Text and Cases* [22] , provide excellent descriptions and examples of socio-technical approaches to information security management in today's organizations. Information security forum's webpages (www.securityforum.org) give examples of practical guidance related to information security management.

Visit the webpages of the European Network and Information Security Agency, Awareness Raising (www.enisa.europa.eu/act/ar) for more practical information on information security awareness. Furthermore, the information security forum at www.securityforum.org offers practical guidance for information security management.

Bibliography

[1] J.S. Nye and J. Donahue, Eds.. *Governance in a Globalized World*. Brookings Institution, Washington DC, 2000.

[2] J. Hovden. *The Development of New Safety Regulations in the Norwegian Oil and Gas Industry*. Ch. 5 in *Changing Regulation. Controlling Risks in Society*. B. Kirwan, A. Hale, and A. Hopkins, Eds., Pergamon, Elsevier Science, Kidlington, Oxford, UK, 2002.

[3] K. McLaughlin, S. P. Obsbourne, and E. Ferlie. *New Public Management. Current Trends and Future Prospects*. Routledge, New York, 2002.

[4] T. Aven. *Foundation of Risk Analysis: A Knowledge and Decision-Oriented Perspective*. Wiley, Chichester, UK, 2002.

[5] O. Renn. *Risk Governance. Coping with Uncertainty in a Complex World*. Earthscan, London, UK, 2008.

[6] OECD. *Emerging Systemic Risks in the 21st Century: An Agenda for Action*. Final Report to OECD Futures Project, Paris, 2003.

[7] E. Hollnagel, D.D. Woods, and N. Leveson. *Resilience Engineering: Concepts and Precepts*. Ashgate, Aldershot, UK, 2006.

[8] G. Dhillon. *Principles of Information Systems Security: Text and Cases*. Wiley, USA, 2007.

[9] J. Rasmussen. Risk management in a dynamic society: A modeling problem. *J. Safety Science*, 27(2–3):183–213, 1997.

[10] U. Beck. *The Risk Society: Towards a New Modernity Sage*, London, UK, 1992.

[11] L. Bogen. *Organisering av IT-sikkerhet i statlig sektor. [Organizing Information Security in the Public Sector]*. MasterS thesis, NTNU, Trondheim, Norway, 2005.

[12] B. Kirwan, A. Hale, and A. Hopkins, Eds. *Changing Regulation: Controlling Risks in Society*. Pergamon, Elsevier Science, Kidlington, Oxford, UK, 2002.

[13] G. Dhillon and J. Backhouse. *Current Directions in IS Security Research: Towards Socio-organizational Perspectives. Information Systems Journal*, 11(2):127–153, 2001.

[14] M.T. Siponen and H. Oinas-Kukkonen. A review of information security issues and respective research contributions. *Database for Advances in Information Systems*, 38(1):60, 2007.

[15] E. Albrechtsen. *Friend or Foe? Information Security Management of Employees*. Doctoral thesis, Norwegian University of Science and Technology, 2008.

[16] E. Trist and K. W. Bamforth. Some social and psychological consequences of the longwall method of coal getting. *Human Relations*, 4(1):3–38, 1951.

[17] E. Trist. *The Evolution of Socio-technical Systems: A Conceptual Framework and an Action Research Program*. Ontario Quality of Working Life Centre, Toronto, 1981.

[18] B. Schneier. *Secrets and Lies : Digital Security in a Networked World*. New York, Wiley, 2000.

[19] H. Mintzberg. *The Structuring of Organizations*. Englewood Cliffs, NJ, Prentice Hall, 1979.

[20] L. G. Bolman and T.E. Deal. *Modern approaches to understanding and managing organizations*. San Francisco, Jossey-Bass, 1984.

[21] G. Dhillon. *Information Security Management. Global Challenges in the New Millennium*. London, Idea Group Publishing, 2001.

[22] G. Dhillon. *Principles of Information Systems Security. Text and Cases*. Wiley, New York, 2007.

[23] E. Albrechtsen and T.O. Grøtan. *Gammeldags tenkning i moderne organisasjoner? Om IKT-sikkerhet i kunnskapsorganisasjoner.* [*Old-fashioned thinking in modern organizations? On ICT security in knowledge organizations*]. In Lydersen (ed.), *Fra is i fingeren til ragnarok: Tjue historier om sikkerhet*, 335–355, 38(1):60, Tapir Akademisk, Trondheim, 2004.

[24] ISO/IEC. 27001:2005: Information technology – Security techniques – Information security management systems – Requirements.

[25] G. Dhillon and J. Backhouse. Information System Security Management in the New Millenium. *Communications of the ACM.* 43(7):125–128, 2000.

[26] G. B. Magklaras and S. M. Furnell. Insider threat prediction tool: Evaluating the probability of IT misuse. *Computers & Security* 21(1):62–73, 2001.

[27] K. D. Mitnick and W. L. Simon. *The Art of Deception: Controlling the Human Element of Security.* Wiley, Indianapolis, 2002.

[28] L. A. Gordon, M. P. Loeb, W. Lucyshyn, and R. Richardson. *2005 CSI/FBI Computer Crime and Security Service.* Computer Security Institute, 2005.

[29] M. E. Whitman. Enemy at the gate: Threats to information security. *Communications of the ACM* 46(8):91–95, 2003.

[30] M. Theoharidou, S. Kokolakis, M. Karyda, and E. Kiountouzis. The insider threat to information systems and the effectiveness of ISO17799. *Computers & Security* 24(6):472–484, 2005.

[31] J. Reason. *Managing the Risks of Organizational Accidents.* Aldershot, Ashgate, 1997.

[32] E. Hollnagel. *Barriers and Accident Prevention.* Aldershot, Ashgate, 2004.

[33] P. M. Schiefloe. *Mennesker og samfunn: Innføring i sosiologisk forståaelse.* In Norwegian. [*Humans and Society: Introduction to Sociology*] Fagbokforl, Bergen, 2003.

[34] T. Rundmo. *Atferdsvitenskaplig sikkerhetsforskning.* In Norwegian. [*Safety research on behaviour*] SINTEF-report no. STF38A01408M, 1990.

[35] J. Hovden, P. Ingstad, B. Mostue, R. Rosness, T. Rundmo, and R. K. Tinmannsvik. *Ulykkesforebyggende arbeid.* In Norwegian. [*Accident Prevention*], Yrkeslitteratur, Oslo, 1992.

[36] E. Albrechtsen and J. M. Hagen. *Information security measures influencing user performance.* In Proceedings of the European Safety and Reliability Conference, 2008.

[37] A. R. Hale and A. I. Glendon. *Individual Behaviour in the Control of Danger.* Elsevier, Amsterdam, 1987.

[38] B. D. Voss. *The Ultimate Defence of Depth: Security Awareness in Your Company.* SANS Institute White Paper, 2001.

[39] W. Hubbard. *Methods and Techniques of Implementing a Security Awareness Program.* SANS Institute White Paper, 2002.

[40] E. Albrechtsen and J. Hovden. *User participation in information security.* In Proceedings of the European Safety and Reliability Conference 2007.

[41] E. H. Schein. *Organizational Culture and Leadership*. San Fransisco, Jossey-Bass, 1992.

[42] R. Westrum. *Cultures with Requisite Imagination*. In J. Wise, D. Hopkins, and P. Stager (eds.), V*erification and Validation of Complex Systems: Human Factors Issues*. Berlin, Springer-Verlag, 1992.

[43] K. Koh, A. B. Ruighaver, S. Maynard, and A. Ahmad. *Security Governance: Its impact on Security culture*. In Proceedings of the 3rd Australian Information Security Management Conference, Perth, 2005.

[44] A. B. Ruighaver, S. B. Maynard, and S. Chang. Organisational security culture: Extending the end-user perspective. *Computers & Security*. 26(1):56–62, 2007.

[45] A. R. Hale. Culture's confusions. *Safety Science*. 34(1–3):1–14, 2000.

Index

Printed and bound by CPI Group (UK) Ltd, Croydon, CR0 4YY

25/10/2024

01779245-0001